An Introduction to Spatial Data Science with GeoDa

Volume 2 – Clustering Spatial Data

This book is the second in a two-volume series that introduces the field of spatial data science. It moves beyond pure data exploration to the organization of observations into meaningful groups, i.e., spatial clustering. This constitutes an important component of so-called unsupervised learning, a major aspect of modern machine learning.

The distinctive aspects of the book are both to explore ways to spatialize classic clustering methods through linked maps and graphs, as well as the explicit introduction of spatial contiguity constraints into clustering algorithms. Leveraging a large number of real-world empirical illustrations, readers will gain an understanding of the main concepts and techniques and their relative advantages and disadvantages. The book also constitutes the definitive user's guide for these methods as implemented in the GeoDa open-source software for spatial analysis.

It is organized into three major parts, dealing with dimension reduction (principal components, multidimensional scaling, stochastic network embedding), classic clustering methods (hierarchical clustering, k-means, k-medians, k-medoids and spectral clustering) and spatially constrained clustering methods (both hierarchical and partitioning). It closes with an assessment of spatial and non-spatial cluster properties.

The book is intended for readers interested in going beyond simple mapping of geographical data to gain insight into interesting patterns as expressed in spatial clusters of observations. Familiarity with the material in Volume 1 is assumed, especially the analysis of local spatial autocorrelation and the full range of visualization methods.

Luc Anselin is the Founding Director of the Center for Spatial Data Science at the University of Chicago, where he is also Stein-Freiler Distinguished Service Professor of Sociology and the College, as well as a member of the Committee on Data Science. He is the creator of the GeoDa software and an active contributor to the PySAL Python open-source software library for spatial analysis. He has written widely on topics dealing with the methodology of spatial data analysis, including his classic 1988 text on Spatial Econometrics. His work has been recognized by many awards, such as his election to the U.S. National Academy of Science and the American Academy of Arts and Science.

An Introduction to Spatial Data Science with GeoDa

Volume 2 – Clustering Spatial Data

Luc Anselin

CRC Press
Taylor & Francis Group
Boca Raton London New York

CRC Press is an imprint of the
Taylor & Francis Group, an **informa** business

A CHAPMAN & HALL BOOK

Designed cover image: © Luc Anselin

First edition published 2024
by CRC Press
2385 NW Executive Center Drive, Suite 320, Boca Raton FL 33431

and by CRC Press
4 Park Square, Milton Park, Abingdon, Oxon, OX14 4RN

CRC Press is an imprint of Taylor & Francis Group, LLC

ISBN: 978-1-032-71302-1 (hbk)
ISBN: 978-1-032-71316-8 (pbk)
ISBN: 978-1-032-71317-5 (ebk)

DOI: 10.1201/9781032713175

Typeset in Latin Modern font
by KnowledgeWorks Global Ltd.

Publisher's note: This book has been prepared from camera-ready copy provided by the authors.

To Emily

Contents

List of Figures

Preface

In contrast to the materials covered in Volume 1, this second volume has no precedent in an earlier workbook. Much of its contents have been added in recent years to the `GeoDa` documentation pages, as the topics were gradually included into my *Introduction to Spatial Data Science* course and implemented in `GeoDa`. At one point, the material became too much to constitute a single course and was split off into a separate *Spatial Clustering* course. The division of the content between the two volumes follows this organization.

In contrast to the first volume, where the focus is almost exclusively on data exploration, here attention switches to the delineation of groupings of observations, i.e., *clusters*. Both traditional and spatially constrained methods are considered. Again, the emphasis is on how a *spatial perspective* can contribute to additional insight, both by considering the spatial aspects explicitly (as in spatially constrained clustering) as well as through *spatializing* classic techniques.

Compared to Volume 1, the treatment is slightly more mathematical and familiarity with the methods covered in the first volume is assumed. As before, extensive references are provided. However, in contrast to the first volume, several methods included here are new and have not been treated extensively in earlier publications. They were typically introduced as part of the documentation of new features in `GeoDa`.

The empirical illustrations use the same sample data sets as in Volume 1. These are included in the software.

All applications are based on Version 1.22 of the software, available in Summer 2023. Later versions may include slight changes as well as additional features, but the treatment provided here should remain valid. The software is free, cross-platform and open source, and can be downloaded from https://geodacenter.github.io/download.html.

Acknowledgments

This second volume is based on enhancements in the `GeoDa` software implemented in the past five or so years, with Xun Li as the lead software engineer and Julia Koschinsky as a constant source of inspiration and constructive comments. The software development received institutional support by the University of Chicago to the Center for Spatial Data Science.

Help and suggestions with the production process from Lara Spieker of Chapman & Hall is greatly appreciated.

As for the first volume, Emily has been patiently living with my GeoDa obsession for many years. This volume is also dedicated to her.

Shelby, MI, Summer 2023

About the Author

Luc Anselin is the Founding Director of the Center for Spatial Data Science at the University of Chicago, where he is also Stein-Freiler Distinguished Service Professor of Sociology and the College. He previously held faculty appointments at Arizona State University, the University of Illinois at Urbana-Champaign, the University of Texas at Dallas, the Regional Research Institute at West Virginia University, the University of California, Santa Barbara, and The Ohio State University. He also was a visiting professor at Brown University and MIT. He holds a PhD in Regional Science from Cornell University.

Over the past four decades, he has developed new methods for exploratory spatial data analysis and spatial econometrics, including the widely used local indicators of spatial autocorrelation. His 1988 *Spatial Econometrics* text has been cited some 17,000 times. He has implemented these methods into software, including the original SpaceStat software, as well as GeoDa, and as part of the Python PySAL library for spatial analysis.

His work has been recognized by several awards, including election to the U.S. National Academy of Sciences and the American Academy of Arts and Sciences.

1

Introduction

This second volume in the *Introduction to Spatial Data Science* is devoted to the topic of spatial clustering. More specifically, it deals with the *grouping* of observations into a smaller number of *clusters*, which are designed to be representative of their members. The techniques considered constitute an important part of so-called *unsupervised learning* in modern machine learning. Purely statistical methods to discover spatial clusters in data are beyond the scope.

In contrast to Volume 1, which assumed very little prior (spatial) knowledge, the current volume is somewhat more advanced. At a minimum, it requires familiarity with the scope of the exploratory toolbox included in the `GeoDa` software. In that sense, it clearly builds upon the material covered in Volume 1. Important principles that are a main part of the discussion in Volume 1 are assumed known. This includes linking and brushing, the various types of maps and graphs, spatial weights and spatial autocorrelation statistics.

Much of the material covered in this volume pertains to methods that have been incorporated into the `GeoDa` software only in the past few years, so as to support the second part of an *Introduction to Spatial Data Science* course sequence. The particular perspective offered is the tight integration of the clustering results with a spatial representation, through customized cluster maps and by exploiting linking and brushing.

The treatment is slightly more technical than in the previous volume, but the mathematical details can readily be skipped if the main interest is in application and interpretation. Necessarily, the discussion relies on somewhat more formal concepts. Some examples are the treatment of matrix eigenvalues and matrix decomposition, the concept of graph Laplacian, essentials of information theory, elements of graph theory, advanced spatial data structures such as quadtree and vantage point tree, and optimization algorithms like gradient search, iterative greedy descent, simulated annealing and tabu search. These concepts are not assumed known but will be explained in the text.

While many of the methods covered constitute part of *mainstream* data science, the perspective offered here is rather unique, with an enduring attempt at *spatializing* the respective methods. In addition, the treatment of *spatially constrained clustering* introduces contiguity as an additional element into clustering algorithms.

Most methods discussed are familiar from the literature, but some are new. Examples include the *common coverage percentage*, a local measure of goodness of fit between distance preserving dimension reduction methods, two new spatial measures to assess cluster quality, i.e., the *join count ratio* and the *cluster match map*, a heuristic to obtain contiguous results from classic clustering results and a hybrid approach toward spatially constrained clustering, whereby the outcome of a given method is used as the initial feasible region in a second method. The techniques are the results of refinements in the software and the presentation of cluster results, and have not been published previously. In addition, the various methods to *spatialize* cluster results are mostly also unique to the treatment in this volume.

As in Volume 1, the coverage here also constitutes the definitive user's guide to the `GeoDa` software, complementing the previous discussion.

In the remainder of this introduction, I provide a broad overview of the organization of Volume 2, followed by a listing of the sample data sets used. As was the case for Volume 1, these data sets are included as part of the `GeoDa` software and do not need to be downloaded separately. For a quick tour of the `GeoDa` software, I refer to the Introduction of Volume 1.

1.1   Overview of Volume 2

Volume 2 is organized into four parts:

- Dimension reduction
- Classic clustering
- Spatial clustering
- Assessment

The first part reviews classic dimension reduction techniques, divided into three chapters, devoted to principal components, multidimensional scaling and stochastic neighbor embedding. In addition to the discussion of the classic properties, specific attention is paid to *spatializing* these techniques, i.e., bringing out interesting spatial aspects of the results.

Part II covers classic clustering methods, in contrast to spatially constrained clusters, which are the topic of Part III. Four chapters deal with, respectively, hierarchical clustering methods, partitioning clustering methods (K-Means), advanced methods (K-Medians and K-Medoids) and spectral clustering.

The chapters in Part III deal with methods to include an explicit spatial constraint of contiguity into the clustering routine. The first chapter outlines techniques to spatialize classic clustering methods, which involve *soft* spatial constraints. These techniques do not guarantee a spatially compact (contiguous) solution. In contrast, the methods discussed in the next two chapters impose *hard* spatial constraints. One chapter deals with hierarchical approaches (spatially constrained hierarchical clustering, SKATER and REDCAP) and the other with partitioning methods (AZP and max-p).

Part IV deals with assessment and includes a final chapter outlining a range of approaches to validate the cluster results, both in terms of internal validity and external validity. It closes with some concluding remarks.

As before, in addition to the material covered in this volume, the GeoDaCenter Github site (https://geodacenter.github.io) contains an extensive support infrastructure. This includes detailed documentation and illustrations, as well as a large collection of sample data sets, cookbook examples and links to a YouTube channel containing lectures and tutorials. Specific software support is provided by means of a list of *frequently asked questions* and *answers to common technical questions*, as well as by the community through the *Google Groups Openspace* list.

1.2 Sample Data Sets

As in Volume 1, the methods and software are illustrated by means of empirical examples that are available directly from inside the GeoDa software. In Volume 2, only a subset of the full slate of sample data are used.

These are

- Italian community banks (n=261)
  - bank performance indicators for 2011-17 (used by Algeri et al., 2022)
  - see Chapters 2 through 4
- Chicago CCA Profiles (n=77)
  - socio-economic snapshot for Chicago Community Areas in 2020 (American Community Survey from the Chicago Metropolitan Agency for Planning – CMAP – data portal)
  - see Chapter 5
- Chicago Census Tracts (n=791)
  - socio-economic determinants of health in 2014 (a subset of the data used in Kolak et al., 2020)
  - see Chapters 6 and 7
- Spirals (n=300)
  - canonical data set to test spectral clustering
  - see Chapter 8
- Municipalities in the State of Ceará, Brazil (n=184)
  - Zika and Microcephaly infections and socio-economic profiles for 2013-2016 (adapted from Amaral et al., 2019)
  - see Chapters 9 through 12

Further details are provided in the context of specific methods.

3.5 Sample Data Sets

Part I

Dimension Reduction

2

Principal Component Analysis (PCA)

The familiar *curse of dimensionality* affects analysis across two dimensions. One is the number of observations (big data) and the other is the number of variables considered. The methods included in Volume 2 address this problem by reducing the dimensionality, either in the number of observations (clustering) or in the number of variables (dimension reduction). The three chapters in Part I address the latter problem. This chapter covers *principal components analysis* (PCA), a core method of both multivariate statistics and machine learning. Dimension reduction is particularly relevant in situations where many variables are available that are highly intercorrelated. In essence, the original variables are replaced by a smaller number of proxies that represent them well in terms of their statistical properties.

Before delving into the formal derivation of principal components, a brief review is included of some basic concepts from matrix algebra, focusing in particular on matrix decomposition. Next follows a discussion of the mathematical properties of principal components and their implementation and interpretation.

A distinct characteristic of this chapter is the attention paid to *spatializing* the inherently non spatial concept of principal components. This is achieved by exploiting geovisualization, linking and brushing to represent the dimension reduction in geographic space. Of particular interest are principal component maps and the connection between univariate local cluster maps for principal components and their multivariate counterpart.

The methods are illustrated using the *Italy Community Banks* sample data set.

2.1 Topics Covered

- Understand the mathematics behind principal component analysis
- Compute principal components for a set of variables
- Interpret the characteristics of a principal component analysis
- Spatialize the principal components
- Investigate the connection between clustering of principal components and multivariate clustering

GeoDa Functions

- Clusters > PCA
    - select variables
    - PCA parameters
    - PCA summary statistics
    - saving PCA results

DOI: 10.1201/9781032713175-2

Toolbar Icons

Figure 2.1: Clusters > PCA | MDS | t-SNE

2.2 Matrix Algebra Review

Before moving on to the mathematics of principal components analysis, a brief review of some basic matrix algebra concepts is included here. Readers already familiar with this material can easily skip this section.

Vectors and matrices are ways to collect a lot of information and manipulate it in a concise mathematical way. One can think of a matrix as a table with rows and columns, and a vector as a single row or column. In two dimensions, i.e., for two values, a vector can be visualized as an arrow between the origin of the coordinate system (0,0) and a given point. The first value corresponds to the x-axis and the second value corresponds to the y-axis. In Figure 2.2, this is illustrated for the vector:

$$v = \begin{bmatrix} 1 \\ 2 \end{bmatrix}.$$

The *arrow* in the figure connects the origin of a $x - y$ scatter plot to the point $(x = 1, y = 2)$.

A central application in matrix algebra is the multiplication of vectors and matrices. The simplest case is the multiplication of a vector by a scalar (i.e., a single number). Graphically, multiplying a vector by a scalar just moves the end point further or closer on the same slope. For example, multiplying the vector v by the scalar 2 gives:

$$2 \times \begin{bmatrix} 1 \\ 2 \end{bmatrix} = \begin{bmatrix} 2 \\ 4 \end{bmatrix}$$

This is equivalent to moving the arrow over on the same slope from (1,2) to the point (2,4) further from the origin, as shown by the dashed red line in Figure 2.2.

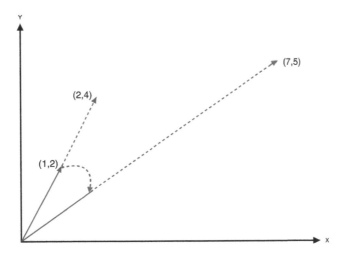

Figure 2.2: Vectors in Two-Dimensional Space

Multiplying a matrix by a vector is slightly more complex, but again corresponds to a simple geometric transformation. For example, consider the 2×2 matrix A:

$$A = \begin{bmatrix} 1 & 3 \\ 3 & 2 \end{bmatrix}.$$

The result of a multiplication of a 2×2 matrix by a 2×1 column vector is a 2×1 column vector. The first element of this vector is obtained as the product of the matching elements of the first row with the vector, the second element similarly as the product of the matching elements of the second row with the vector. In the example, this boils down to:

$$Av = \begin{bmatrix} (1 \times 1) + (3 \times 2) \\ (3 \times 1) + (2 \times 2) \end{bmatrix} = \begin{bmatrix} 7 \\ 5 \end{bmatrix}.$$

Geometrically, this consists of a combination of rescaling and rotation. For example, in Figure 2.2, first the slope of the vector is changed, followed by a rescaling to the point $(7,5)$, as shown by the blue dashed arrows.

A case of particular interest is for any matrix A to find a vector v, such that when post-multiplied by that vector, there is only rescaling and no rotation. In other words, instead of finding what happens to the point $(1,2)$ after pre-multiplying by the matrix A, the interest focuses on finding a particular vector that just moves a point up or down on the same slope for that particular matrix. As it turns out, there are several such solutions. This problem is known as finding *eigenvectors* and *eigenvalues* for a matrix. It has a broad range of applications, including in the computation of principal components.

2.2.1 Eigenvalues and eigenvectors

The eigenvectors and eigenvalues of a square symmetric matrix A are a special scalar-vector pair, such that $Av = \lambda v$, where λ is the *eigenvalue* and v is the *eigenvector*. In addition, the different eigenvectors are such that they are orthogonal to each other. This means that the product of two different eigenvectors is zero, i.e., $v'_u v_k = 0$ (for $u \neq k$).[1] Also, the sum of squares of the eigenvector elements equals one. In vector notation, $v'_u v_u = 1$.

---

[1]The product of a row vector with a column vector is a scalar. The symbol $'$ stands for the transpose of a vector, in this case, a column vector that is turned into a row vector.

What does this mean? For an eigenvector (i.e., arrow from the origin), the transformation by A does not rotate the vector, but simply rescales it (i.e., moves it further or closer to the origin), by exactly the factor λ.

For the example matrix A, the two eigenvectors turn out to be [0.6464 0.7630] and [-0.7630 0.6464], with associated eigenvalues 4.541 and -1.541. Each square matrix has as many eigenvectors and matching eigenvalues as its rank, in this case 2 – for a 2 by 2 nonsingular matrix. The actual computation of eigenvalues and eigenvectors is rather complicated, and is beyond the scope of this discussion.

To further illustrate this concept, consider post-multiplying the matrix A with its eigenvector [0.6464 0.7630]:

$$\begin{bmatrix} (1 \times 0.6464) + (3 \times 0.7630) \\ (3 \times 0.6464) + (2 \times 0.7630) \end{bmatrix} = \begin{bmatrix} 2.935 \\ 3.465 \end{bmatrix}$$

The eigenvector rescaled by the matching eigenvalue gives the same result:

$$4.541 \times \begin{bmatrix} 0.6464 \\ 0.7630 \end{bmatrix} = \begin{bmatrix} 2.935 \\ 3.465 \end{bmatrix}$$

In other words, for the *point* (0.6464 0.7630), a pre-multiplication by the matrix A just moves it by a multiple of 4.541 to a new location on the same slope, without any rotation.

With the eigenvectors stacked in a matrix V, it is easy to verify that they are orthogonal and the sum of squares of the coefficients sum to one, i.e., $V'V = I$ (with I as the identity matrix):

$$\begin{bmatrix} 0.6464 & -0.7630 \\ 0.7630 & 0.6464 \end{bmatrix}' \begin{bmatrix} 0.6464 & -0.7630 \\ 0.7630 & 0.6464 \end{bmatrix} = \begin{bmatrix} 1 & 0 \\ 0 & 1 \end{bmatrix}$$

In addition, it is easily verified that $VV' = I$ as well. This means that the transpose of V is also its inverse (per the definition of an inverse matrix, i.e., a matrix for which the product with the original matrix yields the identity matrix) or $V^{-1} = V'$.

Eigenvectors and eigenvalues are central in many statistical analyses, but it is important to realize they are not as complicated as they may seem at first sight. On the other hand, computing them efficiently *is* complicated, and best left to specialized programs.

Finally, a couple of useful properties of eigenvalues are worth mentioning.

The sum of the eigenvalues equals the *trace* of the matrix. The trace is the sum of the diagonal elements. For the matrix A in the example, the trace is $1 + 2 = 3$. The sum of the two eigenvalues is $4.541 - 1.541 = 3$.

In addition, the product of the eigenvalues equals the *determinant* of the matrix. For a 2×2 matrix, the determinant is $ab - cd$, or the product of the diagonal elements minus the product of the off-diagonal elements. In the example, that is $(1 \times 2) - (3 \times 3) = -7$. The product of the two eigenvalues is $4.541 \times -1.541 = -7.0$.

2.2.2   Matrix decompositions

In many applications in statistics and data science, a lot is gained by representing the original matrix by a product of special matrices, typically related to the eigenvectors and eigenvalues. These are so-called *matrix decompositions*. Two cases are particularly relevant for principal component analysis, as well as many other applications: *spectral decomposition* and *singular value decomposition*.

2.2.2.1 Spectral decomposition

For each eigenvector v of a square symmetric matrix A, it holds that $Av = \lambda v$. This can be written compactly for all the eigenvectors and eigenvalues by organizing the individual eigenvectors as columns in a $k \times k$ matrix V, with k as the dimension of the matrix A. Similarly, the matching eigenvalues can be organized as the diagonal elements of a $k \times k$ diagonal matrix, say G.

The basic eigenvalue expression can then be written as

$$AV = VG.$$

Note that V goes first in the matrix multiplication on the right hand side to ensure that each column of V is multiplied by the corresponding eigenvalue on the diagonal of G to yield λv. Taking advantage of the fact that the eigenvectors are orthogonal, namely that $VV' = I$, gives that post-multiplying each side of the equation by V' yields $AVV' = VGV'$, or

$$A = VGV'.$$

This is the so-called *eigenvalue decomposition* or *spectral decomposition* of the square symmetric matrix A.

For any $n \times k$ matrix of standardized observations X (i.e., n observations on k variables), the square matrix $X'X$ corresponds to the correlation matrix. The spectral decomposition of this matrix yields:

$$X'X = VGV',$$

with V as a matrix with the eigenvectors as columns, and G as a diagonal matrix containing the eigenvalues. This property can be used to construct the principal components of the matrix X.

2.2.2.2 Singular value decomposition (SVD)

The spectral matrix decomposition only applies to square matrices. A more general decomposition is the so-called *singular value decomposition*, which applies to any rectangular matrix, such as the $n \times k$ matrix X with (standardized) observations directly, rather than its correlation matrix.

For a full-rank $n \times k$ matrix X (there are many more general cases where SVD can be applied), the decomposition takes the following form:

$$X = UDV',$$

where U is a $n \times k$ orthonormal matrix (i.e., $U'U = I$), D is a $k \times k$ diagonal matrix and V is a $k \times k$ orthonormal matrix (i.e., $V'V = I$).

While SVD is very general, the full generality is not needed for the PCA case. It turns out that there is a direct connection between the eigenvalues and eigenvectors of the (square) correlation matrix $X'X$ and the SVD of X. Using the SVD decomposition above, and exploiting the orthonormal properties of the various matrices, the product $X'X$ can be written as:

$$X'X = (UDV')'(UDV') = VDU'UDV' = VD^2V',$$

since $U'U = I$.

It thus turns out that the columns of V from the SVD decomposition contain the eigenvectors of $X'X$. In addition, the square of the diagonal elements of D in the SVD are the eigenvalues of the correlation matrix. Or, equivalently, the diagonal elements of the matrix D are the square roots of the eigenvalues of $X'X$. This property can be exploited to derive the principal components of the matrix X.

2.3   Principal Components

Principal components analysis has an eminent historical pedigree, going back to pioneering work in the early twentieth century by the statistician Karl Pearson (Pearson, 1901) and the economist Harold Hotelling (Hotelling, 1933). The technique is also known as the Karhunen-Loève transform in probability theory, and as empirical orthogonal functions or EOF in meteorology (see, for example, in applications of space-time statistics in Cressie and Wikle, 2011; Wikle et al., 2019).

The derivation of principal components can be approached from a number of different perspectives, all leading to the same solution. Perhaps the most common treatment considers the components as the solution of a problem of finding new variables that are constructed as a linear combination of the original variables, such that they maximize the explained variance. In a sense, the principal components can be interpreted as the best linear approximation to the multivariate point cloud of the data.

The point of departure is to organize n observations on k variables x_h, with $h = 1, \ldots, k$, as a $n \times k$ matrix X (each variable is a column in the matrix). In practice, each of the variables is typically standardized, such that its mean is zero and variance equals one. This avoids problems with (large) scale differences between the variables (i.e., some are very small numbers and others very large). For such standardized variables, the $k \times k$ cross product matrix $X'X$ corresponds to the correlation matrix (without standardization, this would be the variance-covariance matrix).[2]

The goal is to find a small number of new variables, the so-called *principal components*, that explain the bulk of the variance (or, in practice, the correlation) in the original variables. If this can be accomplished with a much smaller number of variables than in the original set, the objective of *dimension reduction* will have been achieved.

Each principal component z_u is a linear combination of the original variables x_h, with h going from 1 to k such that:

$$z_u = a_1 x_1 + a_2 x_2 + \cdots + a_k x_k$$

The mathematical problem is to find the coefficients a_h such that the new variables maximize the explained variance of the original variables. In addition, to avoid an indeterminate solution, the coefficients are scaled such that the sum of their squares equals 1.

A full mathematical treatment of the derivation of the optimal solution to this problem is beyond the current scope (for details, see, e.g., Lee and Verleysen, 2007, Chapter 2). Nevertheless, obtaining a basic intuition for the mathematical principles involved is useful.

The coefficients by which the original variables need to be multiplied to obtain each principal component can be shown to correspond to the elements of the eigenvectors of $X'X$, with the

---

[2]The standardization should not be done mechanically, since there are instances where the variance differences between the variables are actually meaningful, e.g., when the scales on which they are measured have a strong substantive meaning (e.g., in psychology).

associated eigenvalue giving the explained variance. Even though the original data matrix X is typically not square (of dimension $n \times k$), the cross-product matrix $X'X$ is of dimension $k \times k$, so it is square and symmetric. As a result, all the eigenvalues are real numbers, which avoids having to deal with complex numbers.

Operationally, the principal component coefficients are obtained by means of a matrix decomposition. One option is to compute the *spectral* decomposition of the $k \times k$ matrix $X'X$, i.e., of the correlation matrix. As shown in Section 2.2.2.1, this yields:

$$X'X = VGV',$$

where V is a $k \times k$ matrix with the eigenvectors as columns (the coefficients needed to construct the principal components) and G a $k \times k$ diagonal matrix of the associated eigenvalues (the explained variance).

A principal component for each observation is obtained by multiplying the row of standardized observations by the column of eigenvalues, i.e., a column of the matrix V. More formally, all the principal components are obtained concisely as:

$$XV.$$

A second and computationally preferred way to approach this is as a *singular value decomposition* (SVD) of the $n \times k$ matrix X, i.e., the matrix of (standardized) observations. From Section 2.2.2.2, this follows as

$$X = UDV',$$

where again V (the transpose of the $k \times k$ matrix V') is the matrix with the eigenvectors of $X'X$ as columns, and D is a $k \times k$ diagonal matrix, containing the square root of the eigenvalues of $X'X$ on the diagonal.[3] Note that the number of eigenvalues used in the spectral decomposition and in SVD is the same, and equals k, the column dimension of X.

Since $V'V = I$, the following result obtains when both sides of the SVD decomposition are post-multiplied by V:

$$XV = UDV'V = UD.$$

In other words, the principal components XV can also be obtained as the product of the orthonormal matrix U with a diagonal matrix containing the square root of the eigenvalues, D. This result is important in the context of multidimensional scaling, considered in Chapter 3.

It turns out that the SVD approach is the solution to viewing the principal components explicitly as a dimension reduction problem, originally considered by Karl Pearson. The observed vector on the k variables x can be expressed as a function of a number of unknown *latent variables* z, such that there is a linear relationship between them:

$$x = Az,$$

where x is a $k \times 1$ vector of the observed variables, and z is a $p \times 1$ vector of the (unobserved) latent variables, ideally with p much smaller than k. The matrix A is of dimension $k \times p$ and contains the coefficients of the transformation. Again, in order to avoid indeterminate solutions, the coefficients are scaled such that $A'A = I$, which ensures that the sum of squares of the coefficients associated with a given component equals one.

---

[3]Since the eigenvalues equal the variance explained by the corresponding component, the diagonal elements of D are thus the standard deviation explained by the component.

Variable	Mean	St.Dev	Min	Max
CAPRAT	0.186	0.068	0.089	0.558
Z	1.286	1.652	0.026	11.128
LIQASS	0.092	0.057	0.020	0.299
NPL	0.146	0.068	0.016	0.333
LLP	0.016	0.011	-0.002	0.056
INTR	0.015	0.004	0.007	0.022
DEPO	0.640	0.095	0.419	0.867
EQLN	0.175	0.073	0.078	0.533
SERV	0.668	0.096	0.373	0.909
EXPE	0.020	0.005	0.009	0.035

Figure 2.3: Italy Bank Characteristics Descriptive Statistics

Instead of maximizing explained variance, the objective is now to find A and z such that the so-called reconstruction error is minimized.[4]

Importantly, different computational approaches to obtain the eigenvalues and eigenvectors (there is no analytical solution) may yield opposite signs for the elements of the eigenvectors. However, the eigenvalues will be the same. The sign of the eigenvectors will affect the sign of the resulting component, i.e., positives become negatives. For example, this can be the difference between results based on a spectral decomposition versus SVD.

In a principal component analysis, the interest typically focuses on three main results. First, the principal component scores are used as a replacement for the original variables. This is particularly relevant when a small number of components explain a substantial share of the original variance. Second, the relative contribution of each of the original variables to each principal component is of interest. Finally, the variance proportion explained by each component in and of itself is also important.

2.3.1   Implementation

Principal components are invoked from the drop-down list created by the toolbar **Clusters** icon (Figure 2.1) as the top item (more precisely, the first item in the dimension reduction category). Alternatively, from the main menu, **Clusters > PCA** gets the process started.

The illustration uses ten variables that characterize the efficiency of community banks, based on the observations for 2013 from the *Italy Community Bank* sample data set (see Algeri et al., 2022):

- CAPRAT: ratio of capital over risk weighted assets
- Z: z score of return on assets (ROA) + leverage over the standard deviation of ROA
- LIQASS: ratio of liquid assets over total assets
- NPL: ratio of non performing loans over total loans
- LLP: ratio of loan loss provision over customer loans
- INTR: ratio of interest expense over total funds
- DEPO: ratio of total deposits over total assets
- EQLN: ratio of total equity over customer loans

---

[4]The concept of reconstruction error is somewhat technical. If A were a square matrix, one could solve for z as $z = A^{-1}x$, where A^{-1} is the inverse of the matrix A. However, due to the dimension reduction, A is not square, so something called a pseudo-inverse or Moore-Penrose inverse must be used. This is the $p \times k$ matrix $(A'A)^{-1}A'$, such that $z = (A'A)^{-1}A'x$. Furthermore, because $A'A = I$, this simplifies to $z = A'x$ (of course, so far the elements of A are unknown). Since $x = Az$, if A were known, x could be found as Az, or, as $AA'x$. The reconstruction error is then the squared difference between x and $AA'x$. The objective is to find the coefficients for A that minimize this expression. For an extensive technical discussion, see Lee and Verleysen (2007), Chapter 2.

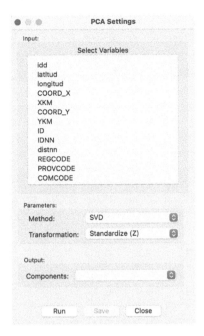

Figure 2.4: PCA Settings Menu

- SERV: ratio of net interest income over total operating revenues
- EXPE: ratio of operating expenses over total assets

Some descriptive statistics are listed in Figure 2.3. An analysis of individual box plots (not shown) reveals that most distributions are quite skewed, with only NPL, INTR and DEPO not showing any outliers. SERV is the only variable with outliers at the low end of the distribution. All the other variables have a considerable number of outlying observations at the high end (see Algeri et al., 2022, for a substantive discussion of the variables).

The correlation matrix (not shown) includes both very strong linear relations between pairs of variables as well as very low ones. For example, NPL is highly correlated with SERV (-0.90) and LLP (0.64), as is CAPRATA with EQLN (0.87), but LIQASS is essentially uncorrelated with NPL (-0.004) and SERV (0.01).

Selection of PCA brings up the **PCA Settings Menu**, which is the main interface to specify all the settings. This interface has a similar structure for all multivariate analyses and is shown in Figure 2.4.

The top dialog is the interface to **Select Variables**. The default **Method** to compute the various coefficients is **SVD**. The other option is **Eigen**, which uses spectral decomposition. By default, all variables are used as **Standardize (Z)**, such that the mean is zero and the standard deviation is one.[5]

The **Run** button computes the principal components and brings up the results in the right-hand panel, as shown in Figure 2.5.

The result summary is evaluated in more detail in Section 2.3.2.

---

[5] A full list of the standardization options in GeoDa is given in Chapter 2 of Volume 1.

```
PCA method: svd

Standard deviation:
1.728221 1.327848 1.123773 0.980447 0.918320 0.859272 0.779411 0.650947 0.550930 0.330976

Proportion of variance:
0.298674 0.176318 0.126286 0.096127 0.084331 0.073835 0.060748 0.042373 0.030352 0.010955

Cumulative proportion:
0.298674 0.474992 0.601279 0.697406 0.781737 0.855572 0.916320 0.958693 0.989046 1.000000

Kaiser criterion: 3.000000

95% threshold criterion: 7.000000

Eigenvalues:
 2.98675
 1.76318
 1.26287
 0.961276
 0.843311
 0.738349
 0.607481
 0.423732
 0.303523
 0.109545

Variable Loadings:
                   PC1          PC2          PC3          PC4          PC5          PC6          PC7          PC8          PC9          PC10
CAPRAT (2013)   0.416651     0.289011     0.400282     0.200753     0.0544673   -0.0346242   -0.177808     0.218361    -0.0207654   -0.677092
Z (2013)        0.0139727   -0.420044     0.235985     0.564699    -0.291502     0.392043     0.451627     0.0676648   -0.0396299   -0.00281511
LIQASS (2013)   0.147682     0.229039    -0.390256     0.638757     0.202746    -0.50688      0.204524    -0.129197     0.000251114  0.0941635
NPL (2013)     -0.264109     0.546288     0.0813843   -0.105872    -0.0697708    0.120595     0.408614    -0.0631704   -0.650065    -0.0321394
LLP (2013)     -0.332622     0.452568     0.221917     0.0941261   -0.124286     0.234403     0.169731    -0.234213     0.687286     0.00442249
INTR (2013)    -0.454442     0.0304158    0.139912     0.081664    -0.0151363   -0.309591     0.0131678    0.807004     0.0830592    0.109132
DEPO (2013)     0.277034     0.133345    -0.483049    -0.0642759    0.381633     0.541986     0.163713     0.443464     0.0677232    0.023698
EQLN (2013)     0.438027     0.240695     0.440113     0.0684679    0.026698     0.0485461   -0.14362      0.0846984   -0.0596709    0.719292
SERV (2013)     0.362998    -0.0664945    0.0908852   -0.443383    -0.113545    -0.349948     0.675579     0.0681695    0.247354    -0.0369921
EXPE (2013)     0.130787     0.321352    -0.347914     0.0256451   -0.831361     0.0653642   -0.151014     0.111648     0.163776     0.019992

Squared correlations:
                   PC1          PC2          PC3          PC4          PC5          PC6          PC7          PC8          PC9          PC10
CAPRAT (2013)   0.518493     0.147273     0.202343     0.0387411    0.00250184   0.000885159  0.0192059    0.020204     0.000130884  0.0502209
Z (2013)        0.000583123  0.311089     0.0703274    0.306536     0.0716588    0.113482     0.123906     0.00194005   0.000476694  8.68711e-07
LIQASS (2013)   0.0651403    0.092494     0.192334     0.392211     0.0346649    0.189701     0.0254109    0.00707289   1.91515e-08  0.000971321
NPL (2013)      0.208336     0.526186     0.0083645    0.0107749    0.0041052    0.0107379    0.101428     0.00169089   0.128264     0.000113153
LLP (2013)      0.330446     0.36113      0.0621927    0.0085166    0.0130266    0.0405682    0.0175008    0.0232441    0.143373     2.14209e-06
INTR (2013)     0.616815     0.00163117   0.0247212    0.00641071   0.00019321   0.0707682    0.000105332  0.275957     0.00209398   0.00130461
DEPO (2013)     0.229227     0.0313509    0.294673     0.00397139   0.122823     0.216889     0.0162817    0.0833313    0.00139208   6.15269e-05
EQLN (2013)     0.57306      0.102148     0.244617     0.00450628   0.000601103  0.00174008   0.0125303    0.00303977   0.00108075   0.0566768
SERV (2013)     0.393556     0.00779589   0.0104314    0.188975     0.0108725    0.0904206    0.277258     0.00196912   0.0185706    0.000149897
EXPE (2013)     0.0510888    0.182078     0.152863     0.000632192  0.582863     0.00315454   0.0138536    0.00528189   0.00814122   4.37872e-05
```

Figure 2.5: PCA Results

	PC1	PC2	PC3	PC4	PC5	PC6	PC7
1	2.262445	0.433308	-3.404982	0.963866	-1.042364	-1.161484	0.081744
2	1.176860	1.974104	-1.420329	2.610017	0.204041	-1.160069	-1.110146
3	-1.516628	0.676118	-1.036056	0.888420	0.977295	0.138653	-0.311785
4	-2.314559	0.926752	-0.191556	0.337600	0.320606	0.203871	-0.303190
5	-2.365019	-0.239757	0.476389	0.330718	0.282252	-0.468738	-0.423377
6	2.543710	-0.089001	-1.305192	2.370842	3.899258	-1.235648	-0.546285
7	-3.468496	3.455145	-0.347688	0.653919	-1.114478	-0.735111	1.070230

Figure 2.6: Principal Components in the Data Table

2.3.1.1 Saving the principal components

Once the computation is finished, the resulting principal components become available to be added to the Data Table as new variables. The **Components** drop-down list suggests the number of components based on the 95% variance criterion (see Section 2.3.2). In the example, this is **7**.

Invoking the **Save** button brings up a dialog to specify the variable names for the principal components, with as default **PC1, PC2**, etc. These variables are added to the Data Table and become available for any subsequent analysis or visualization. This is illustrated in Figure 2.6 for seven components based on the ten original bank variables.

2.3.1.2 Saving the result summary

The listing of the PCA results including eigenvalues, loadings and squared correlations can be saved to a text file by a right-click in the window and selecting **Save**. The resulting text file is an exact replica of the result listing.

2.3.2 Interpretation

The panel with summary results (Figure 2.5) provides several statistics pertaining to the variance decomposition, the eigenvalues, the variable loadings and the contribution of each of the original variables to the respective components.

2.3.2.1 Explained variance

After listing the **PCA method** (here the default **SVD**), the first item in the results panel gives the **Standard deviation** explained by each of the components. It corresponds to the square root of the **Eigenvalues** (each eigenvalue equals the variance explained by the corresponding principal component), which are listed as well. In the example, the first eigenvalue is 2.98675, which is thus the variance of the first component. Consequently, the standard deviation is the square root of this value, i.e., 1.728221, given as the first item in the list.

The sum of all the eigenvalues is 10, which equals the number of variables, or, more precisely, the rank of the matrix $X'X$. Therefore, the proportion of variance explained by the first component is $2.98675/10 = 0.2987$, as reported in the list. Similarly, the proportion explained by the second component is 0.1763, so that the cumulative proportion of the first and second components amounts to $0.2987 + 0.1763 = 0.4750$. In other words, the first two components explain a little less than half of the total variance.

The fraction of the total variance explained is listed both as a separate **Proportion** and as a **Cumulative proportion**. The latter is typically used to choose a cut-off for the number of components. A common convention is to take a threshold of 95%, which would suggest 7 components in the example.

An alternative criterion to select the number of components is the so-called **Kaiser** criterion (Kaiser, 1960), which suggests to take the components for which the eigenvalue exceeds **1**. In the example, this would yield 3 components (they explain slightly more than 60% of the total variance).

The bottom part of the results panel is occupied by two tables that have the original variables as rows and the components as columns.

2.3.2.2 Variable loadings

The first table that relates principal components to the original variables shows the **Variable Loadings**. For each principal component (column), this lists the elements of the corresponding eigenvector. The eigenvectors are standardized such that the sum of the squared coefficients equals one. The elements of the eigenvector are the coefficients by which the (standardized) original variables need to be multiplied to construct each component (see Section 2.3.2.3).

It is important to keep in mind that the signs of the loadings may switch, depending on the algorithm that is used in the computation. However, the absolute value of the coefficients remains the same. In the example, setting **Method** to **Eigen** yields the loadings shown in Figure 2.7.

For PC2, PC6, PC7 and PC9, the signs for the loadings are the opposite of those reported in Figure 2.5. This needs to be kept in mind when interpreting the actual value (and sign) of the components and when using the components as variables in subsequent analyses (e.g., regression) and visualization (see Section 2.4).

Variable Loadings:

	PC1	PC2	PC3	PC4	PC5	PC6	PC7	PC8	PC9	PC10
CAPRAT (2013)	0.416651	-0.289011	0.400282	0.200754	0.054467	0.0346244	0.177808	0.218361	0.0207653	-0.677092
Z (2013)	0.0139727	0.420044	0.235985	0.564699	-0.291502	-0.392044	-0.451626	0.0676647	0.0396299	-0.00281491
LIQASS (2013)	0.147681	-0.229038	-0.390256	0.638757	0.202743	0.50688	-0.204524	-0.129197	-0.000251289	0.0941635
NPL (2013)	-0.264109	-0.546288	0.0813842	-0.105872	-0.0697704	-0.120595	-0.408614	-0.0631707	0.650064	-0.0321393
LLP (2013)	-0.332622	-0.452568	0.221917	0.0941269	0.124286	-0.234403	-0.169731	-0.234213	-0.687286	0.00442248
INTR (2013)	-0.454442	-0.0304157	0.139912	0.0816641	-0.015137	0.309591	-0.0131688	0.807004	-0.0830584	0.109131
DEPO (2013)	0.277034	-0.133345	-0.483049	-0.064276	0.381635	-0.541985	-0.163713	0.443464	-0.0677227	0.0236978
EQLN (2013)	0.438027	-0.240695	0.440113	0.0684684	0.0266983	-0.048546	0.14362	0.0846984	0.0596712	0.719291
EXPE (2013)	0.130787	-0.321352	-0.347914	0.0256441	-0.831361	-0.0653662	0.151014	0.111648	-0.163775	0.019992
SERV (2013)	0.362998	0.0664939	0.0908854	-0.443382	-0.113547	0.349947	-0.675579	0.0681695	-0.247354	-0.0369919

Figure 2.7: Variable Loadings Using the EIGEN Mehtod

	Loading	Obs 1	x*loading	Obs 2	x*loading
s_caprat	0.4167	-0.1421	-0.0592	1.6526	0.6886
s_z	0.0140	-0.5394	-0.0075	-0.6578	-0.0092
s_liqass	0.1477	2.7572	0.4072	3.1531	0.4656
s_npl	-0.2641	-1.1935	0.3152	-0.2235	0.0590
s_llp	-0.3326	-1.2110	0.4028	-0.1311	0.0436
s_intr	-0.4544	-0.8613	0.3914	0.1851	-0.0841
s_depo	0.2770	1.3620	0.3773	0.5562	0.1541
s_eqln	0.4380	-0.4282	-0.1876	0.3826	0.1676
s_expe	0.1308	2.5430	0.3326	1.3062	0.1708
s_serv	0.3630	0.7997	0.2903	-1.3201	-0.4792
PC1			2.2624		1.1769

Figure 2.8: Principal Component Calculation

When the original variables are all standardized, each eigenvector coefficient gives a measure of the relative contribution of a variable to the component in question.

2.3.2.3   Variable loadings and principal components

The detailed computation of the principal components is illustrated for the first two observations on the first component in Figure 2.8. The names of the standardized variables are listed in the leftmost column, followed by the principal component loadings (the values in the PC1 column in Figure 2.5). The next column shows the standardized values for each variable for the first observation. These are multiplied by the loading in column four, with the sum listed at the bottom. The value of 2.2624 matches the entry in the first row under PC1 in Figure 2.6.

Similarly, the value of 1.1769 for the second row under PC1 is obtained at the bottom of column 6. Similar calculations can be carried out to verify the other entries.

2.3.2.4   Substantive interpretation – squared correlation

The interpretation and substantive meaning of the principal components can be a challenge. In *factor analysis*, a number of rotations are applied to clarify the contribution of each variable to the different components. The latter are then imbued with meaning such as "social deprivation," "religious climate," etc. Principal component analysis tends to stay away from this, but nevertheless, it is useful to consider the relative contribution of each variable to the respective components.

The table labeled as **Squared correlations** lists those statistics between each of the original variables in a row and the principal component listed in the column. Each row of the table shows how much of the variance in the original variable is explained by each of the components. As a result, the values in each row sum to one.

More insightful is the analysis of each column, which indicates which variables are important in the computation of the matching component. In the example, **INTR** (61.7%), **EQLN** (57.3%) and **CAPRAT** (51.8%) dominate the contributions to the first principal component. In the second component, the main contributor is **NPL** (52.6%), as well as **Z** (31.1%). This

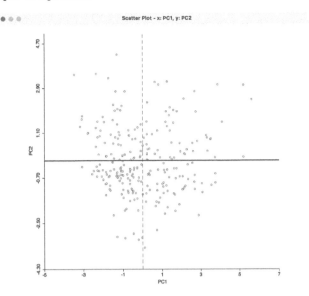

Figure 2.9: Principal Component Scatter Plot

provides a way to interpret how the multiple dimensions along the ten original variables are summarized into the main principal components.

Since the correlations are squared, they do not depend on the sign of the eigenvector elements, unlike the loadings.

2.4 Visualizing principal components

Once the principal components are added to the data table, they are available for analysis in any of the many statistical graphs in `GeoDa` (see Chapters 7–8 in Volume 1). Two use cases warrant some special attention: a scatter plot of any pair of principal components, and the contribution to a component as visualized by means of a parallel coordinate plot.

2.4.1 Scatter plot

A useful graph is a scatter plot of any pair of principal components. For example, such a graph is shown for the first two components (based on the **SVD** method) in Figure 2.9. By construction, the principal component variables are uncorrelated, which yields the characteristic circular cloud plot. A regression line fit to this scatter plot results in a horizontal line (with slope zero). Through linking and brushing, any point or subset of points in the scatter plot can be associated with locations on the map.

In addition, this type of bivariate plot is sometimes employed to visually identify *clusters* in the data. Such clusters would be suggested by distinct high-density clumps of points in the graph. Such points are close in a multivariate sense, since they correspond to a linear combination of the original variables that maximizes explained variance.[6]

---

[6]For an example, see, e.g., Chapter 2 of Everitt et al. (2011).

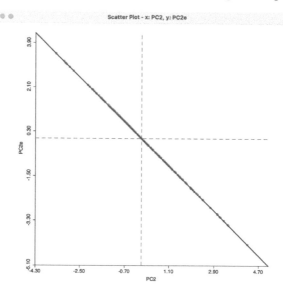

Figure 2.10: SVD versus EIGEN Results

For example, the density cluster methods from Chapter 20 in Volume 1 could be employed to identify clusters among the *points* in the PCA scatter plot. To achieve this, geographical coordinates would be replaced by coordinates along the two principal component dimensions. This provides an alternative perspective on multivariate local clusters.

To illustrate the effect of the choice of eigenvalue computation, Figure 2.10 shows a scatter plot of the second principal component using the **SVD** method (PC2) and the **Eigen** method (PC2e). The sign change is reflected in the perfect negative correlation in the scatter plot.

2.4.2 Multivariate decomposition

Further insight into the connection between a principal component and its constituent variables can be gained from an investigation of a parallel coordinate plot. In the left-hand panel of Figure 2.11, the observations from the bottom 10% of the first principal component are selected (26 observations). The three main contributing variables are shown in the parallel coordinate plot on the right, in standardized form (**s_intr**, **s_eqln**, and **s_caprat**). The selected observations suggest a clustering, with low scores on the principal component corresponding to community banks with a high ratio of interest expenses over total funds, a low ratio of total equity over customer loans and a low ratio of capital over risk-weighted assets, all indicators of rather poor performance.

Again, this illustrates how a univariate principal component can be used as a proxy for multivariate relationships, especially when a high percent of the variance of the original variables is explained, with a distinct and small number of contributing variables to the component.

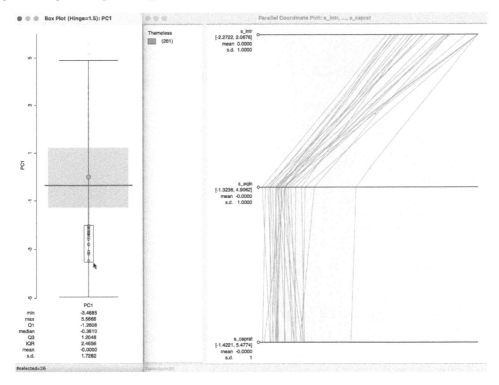

Figure 2.11: Principal Components and PCP

2.5 Spatializing Principal Components

A distinct perspective toward principal component analysis is taken by *spatializing* the visualization, i.e., by directly connecting the value for a principal component to its geographical location. As pointed out before, care must be taken in interpreting the results, since the sign for the component may depend on the method used to compute it. Two special forms of visualization are considered.

One is a thematic map, which may suggest patterns in the values obtained for the components. Mapping of principal components was pioneered in botany and numerical ecology, going back as far as an early paper by Goodall (1954), where principal component scores were mapped using contour lines. A recent review is given in Dray and Jombart (2011).

Dray and Jombart (2011) also consider the *global* spatial autocorrelation among individual principal components, specifically Moran's I, as well as the principal components associated with its extension to multivariate analysis (Dray et al., 2008).[7] However, they do not consider local indicators of spatial autocorrelation. The application of those local cluster methods to principal components provides a univariate alternative to the multivariate cluster analysis considered in Chapter 18 of Volume 1.

---

[7] A related literature pertains to so-called *spatial principal components* or *spatial factors*, but with a focus on global spatial autocorrelation, e.g., Jombart et al. (2008), Frichot et al. (2012).

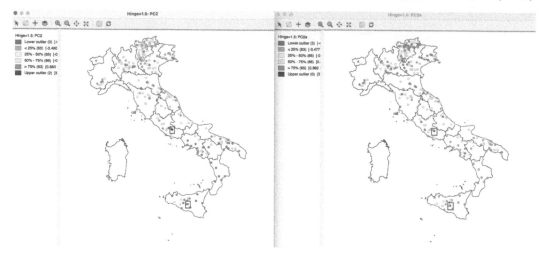

Figure 2.12: Box Map of Second Principal Component

2.5.1   Principal component map

Figure 2.12 illustrates the caution needed when interpreting a map of principal component values. In the left-hand panel, a box map is shown of the second principal component obtained through the SVD method, PC2. On the right is a box map of the same component, but now computed using the Eigen method, PC2e. Clearly, what is high on the left, is low on the right. Specifically, the two upper outliers on the left (observations in the red rectangle) become two lower outliers on the right (observations in the blue rectangle). As a result, what is high or low is less important than the notion of multivariate *similarity*. Observations in the same category share a multivariate similarity that is summarized in the principal component.

2.5.2   Univariate cluster map

A principal component can be treated as any other variable in a local cluster analysis. For example, in the left-hand panel of Figure 2.13, the 19 observations identified as the cores of *Low-Low* clusters are selected, based on the Local Geary statistic, using queen contiguity for the point locations, 99,999 permutations and $p < 0.01$.[8] The matching observations in the parallel coordinate plot on the right illustrate a clustering along multivariate dimensions.

The univariate local cluster map for a principal component can thus be used as a proxy for multivariate clustering of the variables that are the main contributors to the component.

2.5.3   Principal components as multivariate cluster maps

An even closer look at the correspondence between univariate clustering for a principal component and its multivariate counterpart is offered by Figure 2.14. On the left is the same Local Geary cluster map as in Figure 2.13, but now linked to a multivariate Local Geary cluster map for the three main contributing variables (**s_intr**, **s_eqln**, and **s_caprat**). The latter is also based on queen contiguity, with 99,999 permutations and $p < 0.01$. The total number of significant locations in both maps is very similar: 39 in the univariate map and

---

[8]See Chapter 17 in Volume 2 for technical details.

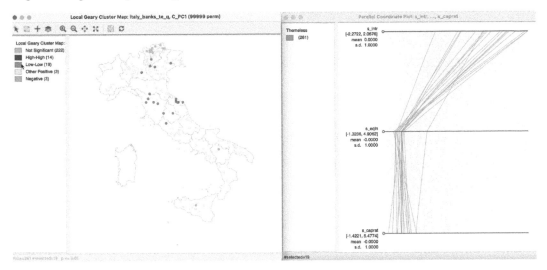

Figure 2.13: Principal Component Local Geary Cluster Map and PCP

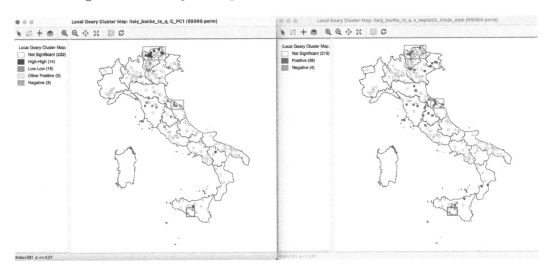

Figure 2.14: Principal Component and Multivariate Local Geary Cluster Map

43 in the multivariate map. Interestingly, the number of spatial outliers is almost identical, with two of them identified for the same locations on the island of Sicily, highlighted by the blue rectangle.

There is also close correspondence between several cluster locations. For example, the *High-High* cluster in the north-east Trentino region and the *Low-Low* cluster in the region of Marche are shared by both maps (highlighted within a green rectangle). While these maps may give similar impressions, it should be noted that in the multivariate Local Geary each variable receives the same weight, whereas the principal component is based on different contributions by each variable.

These findings again suggest that in some instances, a local spatial autocorrelation analysis for one or a few dominant principal components may provide a viable alternative to a full-fledged multivariate analysis. This constitutes a *spatial* aspect of principal components analysis that is absent in standard treatments of the method.

3

Multidimensional Scaling (MDS)

The next two chapters continue the review of approaches to reduce the variable dimension of a problem, but now with a focus on so-called *distance-preserving* methods. Such techniques aim to represent, or, more precisely, *embed* data points from a high-dimensional attribute space in a lower-dimensional space (typically 2D or 3D) for easy visualization. The representation is such that the relative distances between the data points in attribute space are preserved as much as possible.

In this chapter, the topic is *multidimensional scaling* (MDS). Two major methods are considered, one based on an eigenvalue decomposition, the other on *scaling by majorizing a complex function* or **SMACOF**. Each method is presented in some mathematical detail and its implementation is illustrated. A special emphasis continues to be on *spatializing* these methods, by focusing on the connection between attribute space and geographical space. A new measure of the match between spatial and attribute similarity is introduced as the *common coverage percentage*. In addition, the neighbor match test introduced in Volume 1 is extended to MDS solutions.

Some of the discussion is fairly technical and can be readily skipped if the main interest is in application and interpretation.

Again, the *Italy Community Banks* sample data set is used to illustrate these techniques.

3.1 Topics Covered

- Understand the mathematics behind classic metric multidimensional scaling (MDS)
- Understand the principle behind the SMACOF algorithm
- Carry out multidimensional scaling for a set of variables
- Gain insight into the various options used in MDS analysis
- Visualize and interpret the results of MDS
- Compare closeness in attribute space to closeness in geographic space
- Carry out local neighbor match test using MDS neighbors
- Implement density-based clustering on MDS coordinates

GeoDa Functions

- Clusters > MDS
    - select variables
    - MDS methods
    - MDS parameters
    - saving MDS results
    - spatial weights from MDS results

DOI: 10.1201/9781032713175-3

Toolbar Icons

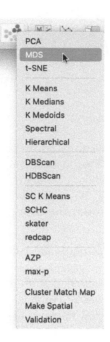

Figure 3.1: Clusters > PCA | MDS | t-SNE

3.2   Classic Metric Scaling

Multidimensional Scaling or MDS is a classic multivariate approach designed to portray the *embedding* of a high-dimensional data cloud in a lower dimension, typically 2D or 3D. The technique goes back to pioneering work on so-called metric scaling by Torgerson (Torgerson, 1952, 1958), and its extension to nonmetric scaling by Shepard and Kruskal (Shepard, 1962a,b; Kruskal, 1964). A very nice overview of the principles and historical evolution of MDS as well as a more complete list of the classic references can be found in Mead (1992).

MDS is based on the notion of distance between observation points in multi-attribute space. For k variables, the *Euclidean* distance between observations x_i and x_j in k-dimensional space is

$$d_{ij} = ||x_i - x_j|| = [\sum_{h=1}^{k}(x_{ih} - x_{jh})^2]^{1/2},$$

the square root of the sum of the squared differences between two data points over each variable dimension.

In this expression, x_i and x_j are k-dimensional column vectors, with one value for each of the k variables (dimensions). In practice, it turns out that working with the squared distance is mathematically more convenient, and it does not affect the properties of the method. In order to differentiate the two notions, the notation d_{ij}^2 will be used for the squared distance.

The squared distance can also be expressed as the inner product of the difference between two vectors:

$$d_{ij}^2 = (x_i - x_j)'(x_i - x_j),$$

with x_i and x_j represented as $k \times 1$ column vectors. It is important to keep the dimensions straight, since the data points are typically contained in an $n \times k$ matrix X, where each observation corresponds to a row of the matrix. However, in order to obtain the sum of squared differences as an inner product, each observation (on the k variables) must be represented as a column vector.

An alternative to Euclidean distance is a *Manhattan* (city block) distance, which is less susceptible to outliers, but does not lend itself to being expressed as an inner product of two vectors. The corresponding expression is:

$$d_{ij} = \sum_{h=1}^{k} |x_{ih} - x_{jh}|,$$

i.e., the sum of the absolute differences in each dimension.

The objective of MDS is to find the coordinates of data points $z_1, z_2, \ldots, z_n$ in 2D or 3D that mimic the distance in multi-attribute space as closely as possible. More precisely, for each observation i ($i = 1, \ldots, n$), z_i consists of a pair of coordinates (for 2D MDS) or a triple (for 3D MDS), which need to be found from the original k-dimensional coordinates.

The problem can be formulated as minimizing a *stress function*, $S(z)$:

$$S(z) = \sum_i \sum_j (d_{ij} - ||z_i - z_j||)^2.$$

In other words, a set of coordinates in a lower dimension (z_i, for $i = 1, \ldots, n$) are found such that the distances between the resulting pairs of points ($||z_i - z_j||$) are as close as possible to their pair-wise distances in high-dimensional multi-attribute space (d_{ij}).

Due to the use of a squared difference, the objective of the stress function penalizes large differentials more. Hence, there is a tendency of MDS to represent points that are far apart rather well, but maybe with less attention to points that are closer together. This may not be the best option when attention focuses on *local* clusters, i.e., groups of observations that are close together in multi-attribute space. An alternative that gives less weight to larger distances is *Stochastic Neighbor Embedding* (SNE), considered in Chapter 4.

3.2.1 Mathematical Details

Classic metric MDS approaches the optimization problem indirectly by focusing on the cross product between the actual vectors of observations x_i and x_j rather than between their difference, i.e., the cross-product between each pair of rows in the observation matrix X. The values for all these cross-products are contained in the matrix expression XX'. This matrix, often referred to as a *Gram* matrix, is of dimension $n \times n$, and not $k \times k$ as is the case for the more familiar $X'X$ used in PCA.

If the actual observation matrix X is available, then the eigenvalue decomposition of the Gram matrix provides the solution to MDS. In the same notation as used for PCA (but now pertaining to the matrix XX'), the spectral decomposition yields:

$$XX' = VGV',$$

where V is the matrix with eigenvectors (each of dimension $n \times 1$) and G is a $n \times n$ diagonal matrix that contains the eigenvalues on the diagonal.

Since the matrix XX' is symmetric, all its eigenvalues are nonnegative, so that their square roots exist. The decomposition can therefore also be written as:

$$XX' = VG^{1/2}G^{1/2}V' = (VG^{1/2})(VG^{1/2})'.$$

In other words, the matrix $VG^{1/2}$ can play the role of X as far as the Gram matrix is concerned. It is not the same as X, in fact it is expressed as coordinates in a different (rotated) axis system, but it yields the same distances (its Gram matrix is the same as XX'). Points in either multidimensional space have the same *relative* positions.

In practice, to express this in lower dimensions, only the first and second (and sometimes third) columns of $VG^{1/2}$ are selected. Consequently, only the two/three largest eigenvalues and corresponding eigenvectors of XX' are required. These can be readily computed by means of the *power iteration method*.

3.2.1.1   Classic metric MDS and principal components

In Chapter 2, the principal components XV were shown to equal the product UD from the SVD of the matrix X. Using the SVD for X, the Gram matrix can also be written as:

$$XX' = (UDV')(VDU') = UD^2U',$$

since $V'V = I$. In other words, the role of $VG^{1/2}$ can also be played by UD, which is the same as the principal components.

More precisely, the first few dimensions of the classic metric MDS will correspond with the first few principal components of the matrix X, associated with the largest eigenvalues (in absolute value). However, the distinguishing characteristic of MDS is that it can be applied without knowing X, as shown in Section 3.2.1.3.

3.2.1.2   Power iteration method

The power method is an iterative approach to obtain the eigenvector associated with the largest eigenvalue of a matrix. It is found from an iteration that starts with an arbitrary unit vector, say x_0.[1] For any given matrix A, the first step is to obtain a new vector as the product of the initial vector and the matrix A, $x_1 = Ax_0$. The procedure iterates by applying higher and higher powers to the transformation $x_k = Ax_{k-1} = A^k x_0$, until convergence (i.e., the difference between x_k and x_{k-1} is less than a pre-defined threshold). The end value, x_k, is a good approximation to the dominant eigenvector of A.[2]

The associated eigenvalue is found from the so-called *Rayleigh* quotient:

$$\lambda = \frac{x_k' A x_k}{x_k' x_k}.$$

The second largest eigenvector is found by applying the same procedure to the matrix B, where B can be viewed as some sort of *residual* after the result for the largest eigenvector has been applied:

$$B = A - \lambda x_k x_k'.$$

The idea can be extended to obtain more eigenvalues/eigenvectors, although typically the power method is only used to compute a few of the largest ones. The procedure can also be applied to the inverse matrix A^{-1} to obtain the smallest eigenvalues.

---

[1]x_0 is typically a vector of ones divided by the square root of the dimension. So, if the dimension is n, the value would be $1/\sqrt{n}$.

[2]For a formal proof, see, e.g., Banerjee and Roy (2014), pp. 353–354.

3.2.1.3 Dissimilarity matrix

The reason why the solution to the metric MDS problem can be found in a fairly straightforward analytic way has to do with the interesting relationship between the squared distance matrix and the Gram matrix for the observations, the matrix XX'. Note that this only holds for Euclidean distances. As a result, the Manhattan distance option is not available for the classic metric MDS.

The squared Euclidean distance between two points i and j in k-dimensional space is:

$$d_{ij}^2 = \sum_{h=1}^{k} (x_{ih} - x_{jh})^2,$$

where x_{ih} corresponds to the h-th value in the i-th row of X (i.e., the i-th observation) and x_{jh} is the same for the j-th row. After working out the individual terms of the squared difference, this becomes:

$$d_{ij}^2 = \sum_{h=1}^{k} (x_{ih}^2 + x_{jh}^2 - 2x_{ih}x_{jh}).$$

The full matrix D^2 for all pairwise distances between observations for a given variable (column) h simply contains the corresponding term in each position $i - j$. It turns out that the squared distance matrix between all pairs of observations for a given variable h can then be written as:

$$D_h^2 = c_h \iota' + \iota c_h' - 2x_h x_h',$$

with $c_h = x_{ih}^2$, the square of the variable h at each location and ι as a $n \times 1$ vector of ones.

The full $n \times n$ squared distance matrix consists of the sum of the individual squared distance matrices over all variables/columns:

$$D^2 = c\iota' + \iota c' - 2XX',$$

since $\sum_{h=1}^{k} x_h x_h' = XX'$, and with each element of $c = \sum_{h=1}^{k} x_{ih}^2$, for $i = 1, \ldots, n$.

This establishes an important connection between the squared distance matrix and the Gram matrix. After some matrix algebra, the relation between the two matrices can be established as:

$$XX' = -\frac{1}{2}(I - M)D^2(I - M),$$

where M is the $n \times n$ matrix $(1/n)\iota\iota'$ and ι is a $n \times 1$ vector of ones (so, M is an $n \times n$ matrix containing $1/n$ in every cell). The matrix $(I - M)$ converts values into deviations from the mean. In the literature, the joint pre- and post-multiplication by $(I - M)$ is referred to as *double centering*.

Given this equivalence, the matrix of observations X is not actually needed (in order to construct XX'), since the eigenvalue decomposition can be applied directly to the double centered squared distance matrix.[3]

3.2.2 Implementation

Classic metric MDS is invoked from the drop-down list created by the toolbar **Clusters** icon (Figure 3.1) as the second item (more precisely, the second item in the dimension reduction category). Alternatively, from the main menu, **Clusters > MDS** gets the process started.

---

[3] A more formal explanation of the mathematical properties can be found in Borg and Groenen (2005), Chapter 7, and Lee and Verleysen (2007), pp. 74–78.

Figure 3.2: MDS Main Dialog

The same ten community bank efficiency characteristics are used as in the treatment of principal components in Chapter 2. They are listed again for ease of interpretation:

- CAPRAT: ratio of capital over risk weighted assets
- Z: z score of return on assets (ROA) + leverage over the standard deviation of ROA
- LIQASS: ratio of liquid assets over total assets
- NPL: ratio of nonperforming loans over total loans
- LLP: ratio of loan loss provision over customer loans
- INTR: ratio of interest expense over total funds
- DEPO: ratio of total deposits over total assets
- EQLN: ratio of total equity over customer loans
- SERV: ratio of net interest income over total operating revenues
- EXPE: ratio of operating expenses over total assets

The process is started with the **MDS Settings** dialog, shown in Figure 3.2. This is where the variables and various options are selected. For now, everything is kept to the default settings, with **Method** as **classic metric** and the **# of Dimensions** set to **2**. The variables are used in standardized form (**Transformation** is set to **Standardize (Z)**), with the usual set of alternative transformations available as options.

3.2.2.1   Saving the MDS coordinates

After the **Run** button is pressed, a small dialog appears to select the variables for the MDS dimensions, i.e., the new x and y axes for a two-dimensional MDS. The default variable names are **V1** and **V2**. With the variable names selected, the new values are added to the data table and the MDS scatter plot is shown, as in Figure 3.3.

Below the scatter plot are two lines with **statistics**. They can be removed by turning the option **View > Statistics** off (the default is on). The variables that were used in the analysis

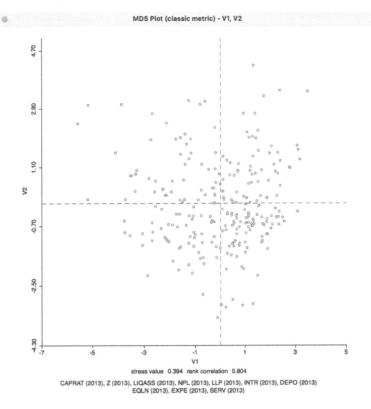

Figure 3.3: MDS Scatter Plot – Default Settings

are listed, as well as two measures of fit, the **stress value** and the **rank correlation**. The stress value gives an indication of how the objective function is met, with smaller values indicating a better fit. The rank correlation is between the original inter-observation distances and the distances in the MDS plot. Here, higher values indicate a better fit. In the example, the results are 0.394 for the stress value and 0.804 for the rank correlation, suggesting a reasonable performance.

3.2.2.2 MDS and PCA

The MDS scatter plot obtained from the classic metric method is essentially the same as the scatter plot of the two first principal components, shown in Figure 2.9. Figure 3.4 indicates how one is the mirror image of the other (flipped around the y-axis). In other words, points in the MDS scatter plot that are close together correspond to points that are close together in the PCA scatter plot.

The equivalence between the two approaches is further illustrated in Figure 3.5, where the values for the x-coordinates in both plots are graphed against each other. There is a perfect negative relationship between the two, i.e., they are identical, except for the sign.

3.2.2.3 Power approximation

An option that is particularly useful in larger data sets is to use a **Power Iteration** to approximate the first few largest eigenvalues needed for the MDS algorithm. Since the first two or three eigenvalue/eigenvector pairs are all that is needed for the classic metric implementation, this provides considerable computational advantages in large data sets.

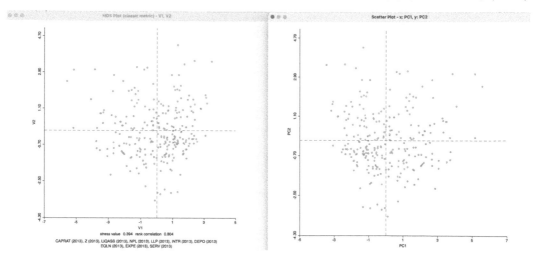

Figure 3.4: MDS and PCA Scatter Plot

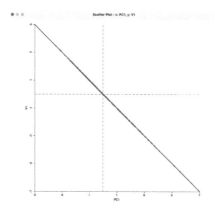

Figure 3.5: MDS Dimensions and PCA

Recall that the matrix for which the eigenvalues/eigenvectors need to be obtained is of dimension $n \times n$. For large n, the standard eigenvalue calculation will tend to break down.

The **Power Iteration** option is selected by checking the corresponding box in the **Parameters** panel of Figure 3.2. This also activates the **Maximum # of Iterations** option. The default is set to **1000**, which should be fine in most situations. However, if needed, it can be adjusted by entering a different value in the corresponding box.

In this example, there is no noticeable difference between the standard solution and the power approximation, as illustrated by the scatter plot in Figure 3.6. However, for data sets with several thousands of observations, the power approximation may offer the only practical option to compute the leading eigenvalues.

Other options to visualize the results of MDS as well as *spatialization* of the MDS analysis are further considered in Sections 3.4 and 3.5.

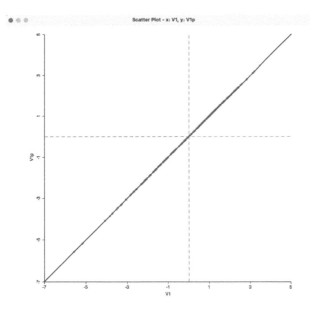

Figure 3.6: MDS Power Approximation

3.3 SMACOF

SMACOF stands for "scaling by majorizing a complex function." The method was initially suggested by de Leeuw (1977) as an alternative to the gradient descent method that was typically employed to minimize the MDS stress function. It was motivated in part because the eigenvalue approach outlined above for classic metric MDS does not work unless the distance function is *Euclidean*. In many early applications of MDS in psychometrics this was not the case.

The idea behind the *iterative majorization* method is actually fairly straightforward, but its implementation can be quite complex. In essence, a *majorizing function* must be found that is much simpler than the original function, and, for a minimization problem, is always *above* the actual function.

With a function $f(x)$, a majorizing function $g(x, z)$ is such that for a fixed point z the two functions are equal, such that $f(z) = g(z, z)$ (z is called the *supporting point*). The auxiliary function should be easy to minimize (easier than $f(x)$) and always dominate the original function, such that $f(x) \leq g(x, z)$. This leads to the so-called *sandwich inequality* (coined as such by de Leeuw):

$$f(x^*) \leq g(x^*, z) \leq g(z, z) = f(z),$$

where x^* is the value that minimizes the function g.

The *iterative* part stands for the way in which one proceeds. One starts with a value x_0, such that $g(x, x_0) = f(x_0)$. For example, consider a parabola that sits above a complex nonlinear function and is tangent to it at x_0, as shown in Figure 3.7.[4] One can easily find the minimum for the parabola, say at x_1.

---

[4]Loosely based on Borg and Groenen (2005), p. 180.

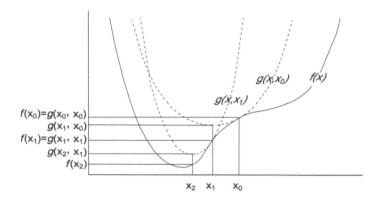

Figure 3.7: Iterative Majorization

Next, a new parabola is moved such that it is tangent to the function at x_1, with again $g(x, x_1) = f(x_1)$. In the following step, the minimum for the new parabola is found at x_2. The procedure continues in this iterative fashion until the difference between x_k and x_{k-1} for two consecutive steps is less than a critical value (the convergence criterion). At that point, the minimum of the parabola is considered to sufficiently approximate the minimum for the function $f(x)$, given a convergence criterion.

3.3.1   Mathematical Details

The application of the SMACOF approach to find the minimum of the stress function is quite complex, and the full technical details are beyond the current scope.[5] The main challenge is to find a suitable majorizing function for the stress function:

$$S(Z) = \sum_{i<j}(\delta_{ij} - d_{ij}(Z))^2 = \sum_{i<j}\delta_{ij}^2 + \sum_{i<j}d_{ij}^2(Z) - 2\sum_{i<j}\delta_{ij}d_{ij}(Z),$$

where Z is a matrix with the *solution* to the MDS problem, δ_{ij} is the distance between i and j for the original configuration and $d_{ij}(Z)$ is the corresponding distance for a proposed solution Z. In general, δ_{ij} can be based on any distance metric or dissimilarity measure, but $d_{ij}(Z)$ has to be a Euclidean distance.

In the stress function, the first term, pertaining to δ_{ij} (the original inter-observation distances), is a constant, since it does not change with the values for the coordinates in Z. The second term is a sum of squared distances between pairs of points in Z, and the third term is a weighted sum of the pairwise distances (each weighted by the initial distances). The objective is to find a set of coordinates Z that minimizes $S(Z)$.

To turn the stress function into a matrix expression suitable for the application of the majorization principle, some special notation is introduced. The difference between observation i and j for a column h in Z, $z_{ih} - z_{jh}$, can be written as $(e_i - e_j)'Z_h$, where e_i and e_j are, respectively, the i-th and j-th column of the identity matrix, and Z_h is the h-th column of Z. As a result, the squared difference between i and j for column h follows as $Z_h'(e_i - e_j)(e_i - e_j)'Z_h$. In the MDS literature, the notation A_{ij} is used for the $n \times n$ matrix

---

[5]For an extensive discussion, see Borg and Groenen (2005), **Chapter 8** and de Leeuw and Mair (2009).

formed by $(e_i - e_j)(e_i - e_j)'$. This is a special matrix consisting of all zeros, except for $a_{ii} = a_{jj} = 1$ and $a_{ij} = a_{ji} = -1$.

Considering all column dimensions of Z jointly then gives the squared distance between i and j as:

$$d_{ij}^2(Z) = \sum_{h=1}^{k} z_h' A_{ij} z_h = tr Z' A_{ij} Z,$$

with tr as the trace operator (the sum of the diagonal elements). Furthermore, summing this over all the pairs $i - j$ (without double counting) gives:

$$\sum_{i<j} d_{ij}^2(Z) = tr(Z' \sum_{i<j} A_{ij} Z) = tr(Z'VZ),$$

with the $n \times n$ matrix $V = \sum_{i<j} A_{ij}$, a row and column-centered matrix (i.e., each row and each column sums to zero), with $n - 1$ on the diagonals and -1 in all other positions. Given the row and column centering, this matrix is singular.

The third term, $-2 \sum_{i<j} \delta_{ij} d_{ij}(Z)$, is the most complex, and the place where the majorization comes into play. Using the same logic as before, it can be written as $-2tr[Z'B(Z)Z]$, where $B(Z)$ is a matrix with off-diagonal elements $B_{ij} = -\delta_{ij}/d_{ij}(Z)$, and diagonal elements $B_{ii} = -\sum_{j,j\neq i} B_{ij}$. In the matrix B, the diagonal elements equal the sum of the absolute values of all the column/row elements, so that rows and columns sum to zero, i.e., the matrix B is double centered.

The majorization is introduced by considering a candidate set of coordinates as Y. After some complex manipulations, a majorization condition follows, which relates the candidate coordinates Y to the solution Z through the so-called *Guttman transform*.[6] This transform expresses an updated solution Z as a function of a tentative solution to the majorizing condition, Y, through the following equality:

$$Z = (1/n)B(Y)Y,$$

where $B(Y)$ has the same structure as $B(Z)$ above, but now using the candidate coordinates from Y.

In practice, the majorization algorithm boils down to an iteration over a number of simple steps:

- start by picking a set of (random) starting values for Z and set Y to these values

- compute the stress function for the current value of Z

- find a new value for Z by means of the Guttman transform, using the computed distances included in $B(Y)$, based on the current value of Y

- update the stress function and check its change; stop when the change is smaller than a pre-set difference

- if convergence is not yet achieved, set Y to the new value of Z and proceed with the update of the stress function

- continue until convergence.

---

[6]Technically, the majorization condition is based on the application of the *Cauchy-Schwarz inequality*, $\sum_m z_m y_m \leq (\sum_m z_m^2)^{1/2} (\sum_m y_m^2)^{1/2}$. Details can be found in Borg and Groenen (2005), Chapter 8.

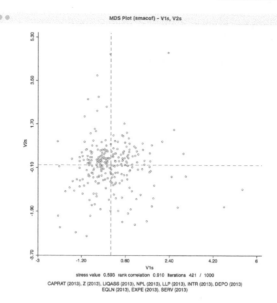

Figure 3.8: MDS Scatter Plot for SMACOF

3.3.2   Implementation

The SMACOF approach is selected by specifying **smacof** as the **Method** in the interface of Figure 3.2. With this selection, the **Maximum # of iterations** and the **Convergence Criterion** options become available (those options are not available to the default **classic metric** method). The default setting is for **1000** iterations and a convergence criterion of **0.000001**. These settings should be fine for most applications, but they may need to be adjusted if the minimization is not accomplished by the maximum number of iterations.

As in the classic metric case, after selecting variable names for the two dimensions (here, **V1s** and **V2s**), a scatter plot is provided, with the method listed in the banner, as in Figure 3.8.

Again, summary statistics of fit are listed at the bottom. In the example, the **stress value** is poorer than the classic metric solution, at 0.593 (relative to 0.394 in Figure 3.3), but the **rank correlation** is better, at 0.910 (relative to 0.084). It is also shown that convergence (for the default criterion) was reached after 421 iterations.

A result of 1000/1000 iterations would indicate that convergence has not been reached. In such an instance, the default number of iterations should be increased. Alternatively, the convergence criterion could be relaxed, but that is not recommended

The overall point pattern of Figure 3.8 is similar to that in Figure 3.3, but it is flipped. A closer investigation of similarities and differences can be carried out by linking and brushing. This will reveal a totally different orientation of the points, with opposite signs for both axes (see also Section 3.3.2.2).

3.3.2.1   Manhattan block distance

With the **smacof** method selected, the **Manhattan** block distance option becomes available as the **Distance Function**. As a result, the original inter-observation distances are based on absolute differences instead of squared differences. The resulting scatter plot (with variable

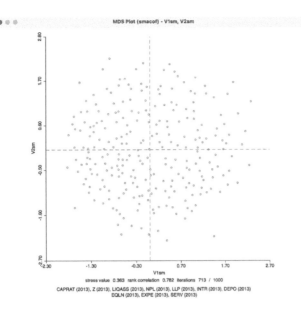

Figure 3.9: MDS Scatter Plot for SMACOF with Manhattan Distance

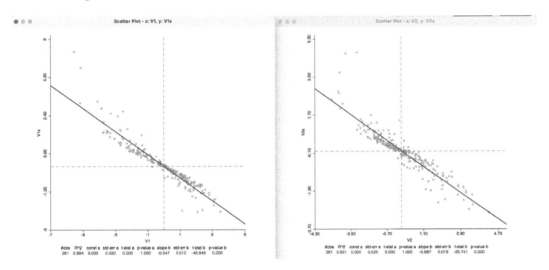

Figure 3.10: MDS Classic vs SMACOF Dimensions

names **V1sm** and **V2sm**) is shown in Figure 3.9. The main effect is to lessen the impact of outliers, or large distances in the original high-dimensional space.

The result that follows from a Manhattan distance metric is quite different from the default Euclidean SMACOF plot in Figure 3.8. Importantly, the scale of the coordinates is not the same, such that the range of values on both axes is much smaller than in the Euclidean case. As a result, even if the points look farther apart, they are actually quite close on the scale of the original plot. However, the identification of *close* observations can differ considerably between the two plots. This can be further explored using linking and brushing.

In terms of fit, the stress value of 0.363 is better than for SMACOF with Euclidean distance, and even better than classic metric MDS, but the rank correlation of 0.782 is worse.

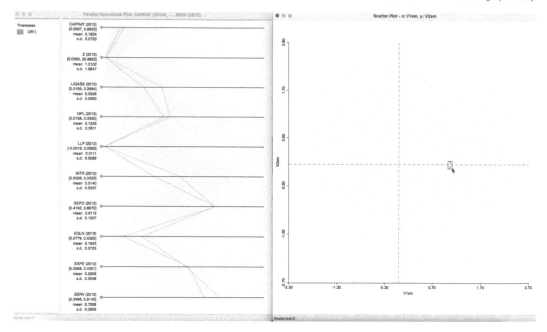

Figure 3.11: MDS and Parallel Coordinate Plot

Convergence is obtained after 713 iterations, substantially more than for the Euclidean distance option.

Overall, this means that the various MDS scatter plots should not be interpreted in a mechanical way, but careful linking and brushing should be employed to explore the trade-offs between the different options.

3.3.2.2 SMACOF vs classic metric MDS

A more formal comparison of the differences between the results for classic metric MDS and SMACOF (both for Euclidean distance) is provided by the scatter plots in Figure 3.10. On the left is a scatter plot that compares the x-axis, on the left one for the y-axis.

Clearly, the signs are opposite. While there is a close linear fit, with an R^2 of 0.894 for the x-axis and 0.831 for the y-axis, there are also some substantial discrepancies, especially at the extreme values.

The advantage of the classic metric method is that its result has a direct relationship with principal components. On the other hand, the SMACOF method allows for the use of a Manhattan distance metric, which is less susceptible to the effect of observations that are far apart in the original multi-attribute space.

3.4 Visualizing MDS

In addition to being integrated in a linking and brushing operation, as suggested in the previous discussion, the results of an MDS analysis can be visualized in a number of different ways. The main insight provided by the MDS scatter plot is the identification of observations

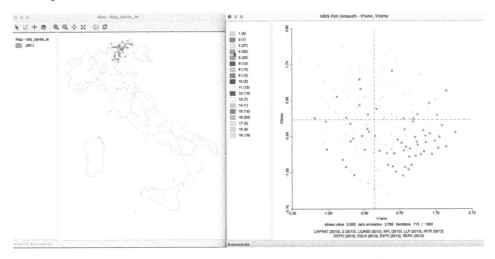

Figure 3.12: MDS with a Categorical Variable

that are close in multi-attribute space, summarizing the information contained in multiple variables down to two (or three) dimensions.

Two particularly insightful nonspatial visualizations can be considered. First, closeness in multi-attribute space can be highlighted by connecting an MDS scatter plot with a parallel coordinate plot. Also, a categorical variable can be introduced into the scatter plot, in a manner similar to the bubble chart discussed in Chapter 8 of Volume 1. For the sake of completeness, the 3D MDS scatter plot is briefly illustrated as well.

3.4.1 MDS and Parallel Coordinate Plot

Figure 3.11 highlights the PCP trajectory over the ten variables used in the analysis for two observations identified as *close* in the MDS scatter plot. The latter, shown in the right-hand panel, is the result of SMACOF for Manhattan distance, the same as Figure 3.9. The trajectories are very close and obtain (near) identical values for five out of the ten variables, with somewhat larger discrepancies for the others. The close pattern illustrates how the two-dimensional graph summarizes a pattern observed in high-dimensional attribute space.

3.4.2 MDS Scatter Plot with Categories

The **MDS Settings** dialog in Figure 3.2 contains a **Category Variable** option. With this option checked and an integer (category) variable selected, the MDS scatter plot turns into a type of bubble chart, with the scatter plot points taking on the color of the category to which they belong. This provides a way to assess whether observations that belong to the same category are also close in multi-attribute space, as summarized by their locations in the MDS scatter plot. In Figure 3.12, the category variable selected is **REGCODE**, the code for the region in which a community bank is located.

The plot on the right shows the points selected for the region of Trentino-Alto Adige (South Tirol), also highlighted in the map on the left. While some of the observations are close together, others are spread out over the scatter plot (again, a SMACOF MDS using Manhattan distance).

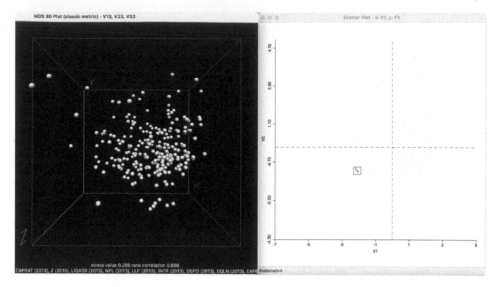

Figure 3.13: 3D MDS

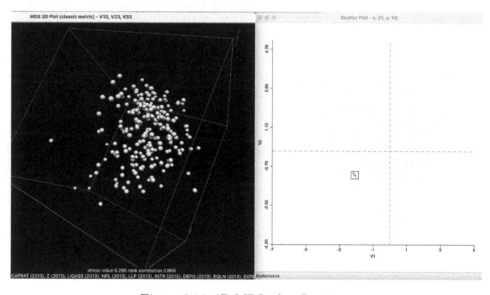

Figure 3.14: 3D MDS after Rotation

Different categorical variables can be considered in this manner (both spatial as well as nonspatial) to assess the extent to which the categories correspond to groupings in multi-attribute space.

3.4.3   3-D MDS

The default setting for MDS analysis is to create a two-dimensional scatter plot. However, the option also exists to obtain results for three dimensions, by specifying **3** for the **# of Dimensions** option. The outcome for classic metric scaling is shown in Figure 3.13, side by side with the two-dimensional scatter plot (same as Figure 3.3). The variables under consideration are listed at the bottom of the graph, together with the stress value (0.295) and the rank correlation (0.866). Both measures of fit are better than in the two-dimensional case, which is not surprising, since less of a dimension reduction is involved.

At first sight, it would seem that the four *close* observations selected in the two-dimensional plot are also close in the 3D scatter plot. However, the perspective used to visualize the three-dimensional volume onto a flat surface can be misleading and the closeness can only be assessed after further manipulation, such as rotation and zooming in.

For example, the rotation depicted in Figure 3.14 illustrates how a change in perspective can reveal larger distances in three dimensions that do not appear in the two-dimensional solution.

3.5 Spatializing MDS

A distinct aspect of the treatment of multi-dimensional scaling in this book, and a perspective largely absent in the rest of the literature, is a focus on *spatializing* the results. More precisely, several approaches are described that allow to investigate a closer link between distance in attribute space and distance in geographical space. This provides an alternative way to approach the problem of multivariate spatial clustering, in which the explicit multivariate aspect is sidestepped by exploiting the relative locations in the MDS scatter plot.

The points in the MDS scatter plot can be viewed as *locations* in a two-dimensional attribute space. This is an example of the use of geographical concepts (location, distance) in contexts where the space is nongeographical. It also allows a further investigation and visualization of the tension between *attribute similarity* and *locational similarity*, two core components underlying the notion of spatial autocorrelation.

An explicit *spatial* perspective is introduced by linking the MDS scatter plot with various map views of the data. In addition, it is possible to exploit the location of the points in attribute space to construct spatial weights based on neighbors in the MDS plot. These weights can then be compared to their geographical counterparts to discover overlap. Furthermore, the point locations themselves can be incorporated in a cluster analysis.

The obvious connection between the MDS scatter plot and a map is explored first, through linking and brushing. Next, the focus shifts to the use of the spatial weights concept to formalize the neighbor structure obtained in the solution of MDS. This is further explored through an alternative to the local neighbor match test of Chapter 18 in Volume 1, based on the k-nearest neighbor structure in the MDS scatter plot. Finally, the application to point locations in the MDS of density-based cluster methods from Chapter 20 in Volume 1 is considered as well.

3.5.1 MDS and Map

An example of linking observations between the MDS scatter plot and a map was given in Figure 3.12, where a subset of observations from the scatter plot was also shown on a map. Instead of using locations only, as in the point map in Figure 3.12, one can also link and brush a thematic map. For example, in Algeri et al. (2022), the bank efficiency indicators are related to an overall technical output efficiency measure.[7] Linking a thematic map for

---

[7] This is a so-called Farrel measure of technical output efficiency based on data envelopment analysis (DEA) under variable returns to scale. For details, see Algeri et al. (2022).

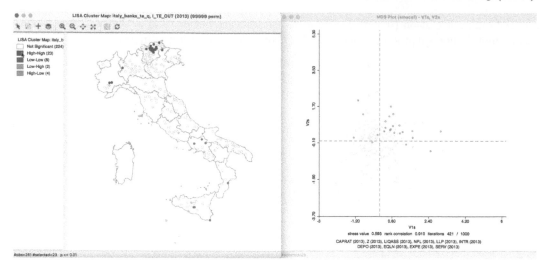

Figure 3.15: Linked Local Moran Cluster Map and MDS

this variable to the MDS scatter plot provides insight into the match of patterns in the map for this *dependent* variable and interesting groupings of the *explanatory* variable in the MDS scatter plot.

Other maps can be linked as well. For example, Figure 3.15 shows a Local Moran cluster map for the technical output efficiency variable, **TE_OUT(2013)** (based on queen contiguity, 99,999 permutations and p < 0.01) linked to the MDS scatter plot for SMACOF, using Euclidean distance (the same as Figure 3.8). The observations in the High-High clusters are selected on the left, and their corresponding points in the MDS space for the explanatory variables highlighted on the right. While several cluster locations have closely aligned points in the MDS, for others this is not the case at all. A careful brushing of the map (and/or other maps) can shed further light on the degree of agreement of the two concepts of clustering.

3.5.2   MDS Spatial Weights

As mentioned, the points in the MDS scatter plot can be viewed as *locations* in an embedded attribute space. As such, they can be *mapped*. In such a point map, the neighbor structure among the points can be exploited to create spatial weights, in exactly the same way as for geographic points (e.g., the distance bands, k-nearest neighbors, contiguity from Thiessen polygons considered in Chapter 11 of Volume 1). Conceptually, such spatial weights are similar to the distance weights created from multiple variables, but they are based on inter-observation distances from an embedded high-dimensional object in two dimensions. While this involves some loss of information, the associated two-dimensional visualization is highly intuitive.

In `GeoDa`, there are three ways to create spatial weights from points in an MDS scatter plot. One is to explicitly create a point layer using the MDS coordinates (e.g., **V1s** and **V2s**), by means of **Tools > Shape > Points from Table**. Once the point layer is in place, the standard spatial weights functionality can be invoked.

A second way pertains only to distance weights. It again uses the **Weights Manager**, but with the **Variables** option for **Distance Weight**, and by specifying the MDS coordinates as the variables. This limits the weights to distance band and k-nearest neighbors. Since

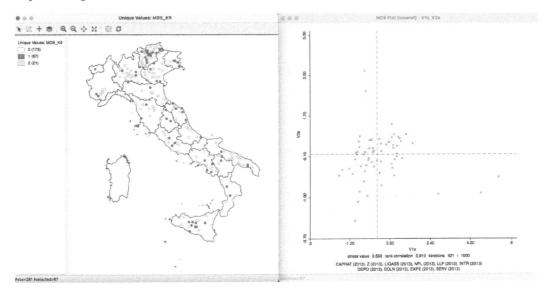

Figure 3.16: Common Connectivity K-6 Neighbors

these coordinates are already normalized in a way, the **Raw** setting is appropriate for standardization.

A final method applies directly to the MDS scatterplot. A right-click in the plot reveals **Create Weights** as the top option.

This brings up the standard **Weights File Creation** interface. The difference with the previous approach is that the MDS coordinates do not need to be specified, but are taken directly from the MDS scatter plot where the option is invoked. In addition, it is also possible to create contiguity weights from the Thiessen polygons associated with the scatter plot points.

3.5.2.1 Attribute and geographical neighbors

An obvious comparison of attribute and geographical similarity is to investigate the extent to which the k-nearest neighbors in geographic space match neighbors in attribute space using the **Intersection** functionality in the **Weights Manager**.

For example, k-nearest neighbor weights for k = 6 and using the SMACOF Euclidean distance can be created using one of the three methods just outlined (i.e., based on **V1s** and **V2s** as the coordinate values). These k-nearest neighbor weights can be intersected with their geographic k-nearest neighbor counterparts (e.g., the file *italy_banks_te_k6.gwt* created in Chapter Chapter 11 of Volume 1). This results in a **gal** weights file that lists for each observation the neighbors that are shared between the two k-nearest neighbor files.

As is to be expected, the resulting file is much sparser than the original weights. The associated **Connectivity Histogram**, reveals that 21 observations have 2 matches, 67 observations have one match, and the remaining 173 observations have no match.

The **Save Connectivity to Table** option can be used to create a variable that reflects this cardinality. Such a variable then lends itself to a **Unique Values Map**, as shown in the left-hand panel of Figure 3.16. The observations with 2 matches are linked to the corresponding points in the MDS scatter plot. This reveals how several of the close neighbors in space are also close neighbors in multi-attribute space, suggesting the presence of *multivariate spatial clustering*.

This is pursued more formally by means of the Local Neighbor Match Test in Section 3.5.3.

3.5.2.2   Common coverage percentage

The **Intersection** functionality for spatial weights can also be employed to compare the nearest neighbor structures implied by different MDS solutions, both with respect to the geographical counterpart as well as for the overlapping neighbor structures between them.

One indication of a different degree of match between attribute and geographical similarity is to compare the percentage nonzero weights in the overlap to the maximum, based on k neighbors for each observation. In this example, the percent nonzero weights for the base result is 2.30%. A perfect match between attribute neighbors and geographic neighbors would therefore yield a ratio of 100%. Various nonspatial weights can be compared to the spatial counterpart by computing the ratio of nonzero weights for the intersection to the maximum possible. This metric can then be referred to as the *common coverage percentage*.

A comparison of the three MDS solutions in terms of overlap with corresponding geographical neighbors yields 0.14% nonzero for the classic metric solution, 0.16% for the SMACOF solution with Euclidean weights, and 0.16% for SMACOF with Manhattan distance. This yields common coverage percentages of respectively 6.09% and 6.96%, showing very similar (low) overlap with spatial neighbors.

This same approach can also be employed to assess the commonality between k-nearest neighbor graphs constructed for different sets of MDS coordinates. In the example, the common coverage percentage is 1.07/2.30 or 46.5% between classic metric and Euclidean SMACOF, 0.08/2.30 or 3.48% between classic metric and Manhattan SMACOF and 1.40/2.30 or 60.87% between the two SMACOF distance metrics. This approach provides a more quantitative assessment of match between MDS solutions than a visual inspection of the MDS scatter plot and highlights the important differences between classic metric and SMACOF using the Manhattan metric.

3.5.3   MDS Neighbor Match Test

A final, more formal approach to check the similarity between neighbors in attribute space and neighbors in geographic space is to consider it as a special case of the local neighbor match test. Instead of using the k-nearest neighbor relation for the high-dimensional space, the same relation can be used in the two- or three-dimensional MDS space.

The accomplish this, the MDS coordinates are selected as the variables in the interface for **Space > Local Neighbor Match Test** (see Chapter 11 of Volume 1). For example, this can be applied to the same **V1s** and **V2s** coordinates as before. As usual, an option is provided to save the cardinality of the weights intersection between the attribute knn weights and the geographical knn weights, as well as the associated probability of finding that many neighbors in common.

The associated unique values map with the common connection shown as a graph is given in Figure 3.17, which, except for the connections, is identical to the left-hand panel in Figure 3.16.

The **cpval** column in the data table reveals that locations with two matches have an associated p-value of 0.0062. However, the p-value associated with one common match is only 0.13, which cannot be deemed to be significant.

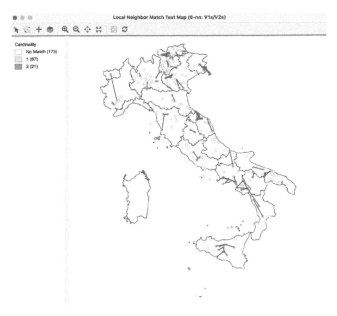

Figure 3.17: Local Neighbor Match Test for MDS

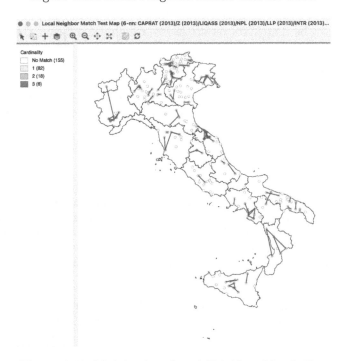

Figure 3.18: Multivariate Local Neighbor Match Test

These results can be compared to a multivariate Local Neighbor Match Test applied to all ten variables. Figure 3.18 shows the results for the latter. Relative to classic MDS, there are more matches for the full multi-attribute result, with six observations achieving three common neighbors (p = 0.000133), in addition to 18 with two neighbors and 82 with one neighbor.

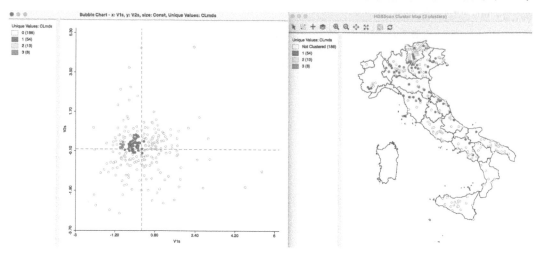

Figure 3.19: HDBSCAN for MDS

An advantage of using the coordinates in embedded space is that the curse of dimensionality is avoided. Instead of having to compute k-nearest neighbors in high dimensions, the MDS results are limited to two or three dimensions, which is computationally easy. In this example, the trade-off seems to be that MDS results in fewer highly significant clusters.

3.5.4   HDBSCAN and MDS

A final *spatial* perspective on the MDS results considered here is the application of density based clustering methods to the MDS coordinate points. This operates in the standard way, as described in Chapter 20 of Volume 1. The clustering routines identify subsets of points in the MDS scatter plot that are *close* together. Indirectly, this provides a way to cluster in multi-attribute space.

Because the density-based clustering results are linked to a map, the extent of geographic association between the cluster points is visualized as well. For example, the right-hand panel of Figure 3.19 shows the outcome of an HDBSCAN routine on the coordinates **V1s** and **V2s**, with a minimum cluster size of **7**. This yields three clusters, containing respectively 54, 13 and 8 observations, depicted in the map. As in the previous analyses, there is some degree of overlap between geographic neighbors and MDS neighbors, but many MDS neighbors are also not geographically close.

The left-hand panel of Figure 3.19 identifies the three clusters on the MDS scatter plot. This figure is created as a bubble chart for a categorical variable, with the cluster category as the variable associated with the bubble color. The three groupings of points are clearly delineated.

The methods outlined in this section provide several alternatives to a full multivariate clustering exercise by means of the summary information contained in the MDS results. In some instances, they may alleviate problems associated with the curse of dimensionality. Together with graphs such as a parallel coordinate plot, further insight can be gained into the relative contribution of variables to the cluster results. By using a combination of these methods, *interesting* locations can be identified and further investigated by means of a careful sensitivity analysis.

4

Stochastic Neighbor Embedding (SNE)

As mentioned, a drawback of the MDS approach is that it employs squared distances in the objective function and thus emphasizes the impact of points that are far apart. As a consequence, MDS tends to do less well to keep the low-dimensional representation of close data points in high dimension also close together in the embedded space. A different approach is offered by *stochastic neighbor embedding* (SNE) (Hinton and Roweis, 2003). Instead of basing the alignment between high and low-dimensional representations on a stress function that includes the actual distances, the problem is reformulated as one of matching two different probability distributions. These distributions replace the explicit distance measure by a probability that any point j is a neighbor of a given point i. This probability decreases with increasing distance between i and j.

In other words, the distance metric of MDS is replaced by a probability measure. The essence of SNE then boils down to matching the probability distribution in high-dimensional space to a distribution in low-dimensional *embedded* space, similar to how distances were matched in MDS.

In this chapter, the basic principles behind SNE and its most recent implementation, *t-SNE* (van der Maaten and Hinton, 2008), are outlined. This material is quite technical. The formal discussion can be readily skipped if the main interest is in application and interpretation.

The chapter starts with a brief introduction to some information-theoretic concepts. This is followed by a description of t-SNE and its implementation. A brief section discusses interpretation and *spatialization*, most of which were covered in detail in Chapter 3 and will not be repeated here. The chapter closes with a comparison of t-SNE to MDS using the common coverage percentage introduced in the previous chapter.

The *Italy Community Banks* sample data set is again used to illustrate these techniques.

4.1 Topics Covered

- Become familiar with fundamental information-theoretic concepts
- Understand the mathematical principles behind t-SNE
- Carry out t-SNE to visualize multi-attribute data
- Understand the role of the various t-SNE optimization parameters
- Interpret the results of t-SNE
- Visualize and spatialize t-SNE results
- Compare t-SNE to MDS

GeoDa Functions

- Clusters > t-SNE
  - select variables

DOI: 10.1201/9781032713175-4

  – t-SNE optimization parameters
  – saving t-SNE results
  – spatial weights from t-SNE results

Toolbar Icons

Figure 4.1: Clusters > PCA | MDS | t-SNE

4.2   Basics of Information Theory

Two important concepts from information theory are part of the SNE logic. The first is the notion of *Shannon entropy*, a measure of the information content of a distribution, the second is a measure of *fit*, the *Kullback-Leibler Information Criterion*, or *KLIC*.

Consider a finite number of events i, each with an associated (discrete) probability p_i. The *entropy* of this distribution is defined as:

$$H = -\sum p_i \ln(p_i).$$

The *maximum* entropy is obtained for the most *diverse* distribution, i.e., a distribution where each event has equal probability. For example, with four equally likely events, $p_i = 0.25$ for each, and the entropy follows as $H = -4 \times (0.25)(-1.386) = 1.386$. This is the highest possible value for four events. One way to view this concept is to think of it as a measure of difficulty to predict a result, e.g., if the four events were cards to be drawn from a deck. With equal probability, it is the hardest to guess which card will be drawn. The least entropy occurs when all the probability is concentrated in one event, which yields $H = 0$. This is also the least diverse distribution.

To make this concept more concrete, consider two more distributions for the four events: one with probabilities 0.125, 0.125, 0.25 and 0.5; and one with probabilities 0.125, 0.125, 0.125

and 0.625. Straightforward calculations yield the entropy for the former as 1.213 and for the latter as 1.074. Because more probability is concentrated in a single event in the latter case, it has a smaller entropy.

In the context of SNE, entropy is rescaled into a measure of *perplexity*, as 2^H. For the three examples considered, the corresponding perplexity is 2.614, 2.318, and 2.105. Perplexity is used as a measure of *how far* one needs to move from a given point to find a sufficient number of neighbors, since the probability becomes a proxy for the distance between two points (i.e., what is the probability that they are neighbors). The lower the entropy, the more high(er) probability is concentrated in fewer events, which means one needs to go less far (smaller bandwidth) to find a desired number of neighbors. Indirectly, therefore, perplexity becomes a proxy for the bandwidth required.

The *Kullback-Leibler* divergence between two distributions is a measure of *fit*, with smaller values indicating a closer correspondence. With two sets of probabilities p_i and q_i, it is defined as:

$$KLIC(P\|Q) = \sum_i p_i \ln \frac{p_i}{q_i} = \sum_i [p_i \ln p_i - p_i \ln q_i].$$

In the example, the KLIC between the equal probability distribution and the second case is 0.173, and between the first and third is 0.291. The measure is not necessarily symmetric. For example, when the role of the first and third distributions are interchanged, the associated KLIC is 0.313.

The KLIC can be interpreted as the difference between the entropy of a distribution and a mixture, where the role of $\ln p_i$ is taken by $\ln q_i$ from the second distribution.

In SNE, the KLIC will be used in the objective function that minimizes the difference in fit between the probabilities in the original high dimensional space (p) and the associated probabilities in the low-dimensional, embedded space (q).

4.2.1   Stochastic Neighbors

In SNE, the concept of multi-attribute distance is replaced by a set of conditional probabilities. More precisely, a conditional probability is introduced for the original high-dimensional data of $p(j|i)$ to express the probability that j is picked as a neighbor of i. In other words, the distance between i and j is converted into a probability measure. Operationally, this is implemented by means of a *kernel* function centered on i. The kernel is a symmetric distribution such that the probability decreases with distance.

For example, using a normal (Gaussian) distribution centered at i with a variance of σ_i^2 would yield the conditional distribution as:

$$p_{j|i} = \frac{exp(-d(x_i, x_j)^2/\sigma_i^2)}{\sum_{h \neq i} exp(-d(x_i, x_h)^2/\sigma_i^2)},$$

with $d(x_i, x_j) = \|x_i - x_j\|$, i.e., the Euclidean distance between i and j in high-dimensional space, and, by convention, $p_{i|i} = 0$. The variance σ_i^2 determines the *bandwidth* of the kernel. It is chosen optimally for each point, such that a target *perplexity* is reached. While a complex concept, the perplexity intuitively relates to a smooth measure of the effective number of neighbors (fewer neighbors relates to a smaller perplexity). In regions with dense points, the variance σ_i^2 should be smaller, whereas in areas with sparse points, it should be larger to reach the same number of neighbors. This is similar in spirit to selecting an adaptive bandwidth based on the k nearest neighbors.

4.3   t-SNE

The t-SNE method, due to van der Maaten and Hinton (2008), is a refinement of the original SNE in two main respects: the conditional distribution is replaced by a joint distribution on (i, j), and the Gaussian kernel is replaced by a thick-tailed normalized Student-t kernel with a single degree of freedom.

The point of departure is a joint probability P for the points in high-dimensional multi-attribute space obtained from the original conditional probabilities as:

$$p_{ij} = \frac{p(j|i) + p(i|j)}{2n},$$

where n is the number of observations.

The counterpart of the distribution P for the high-dimensional points is a distribution Q for the points in a lower dimension. It is defined as:

$$q_{ij} = \frac{(1 + ||z_i - z_j||^2)^{-1}}{\sum_{h \neq l}(1 + ||z_h - z_l||^2)^{-1}},$$

where, as before, the z_i are the coordinates in the embedded (low dimensional) space. The denominator is simply a scaling factor to ensure that the individual joint probabilities sum to one. The numerator is inversely dependent on the Euclidean distance between the points z_i and z_j. In other words, the closer i and j are in embedded space, the higher the associated probability q_{ij} and vice versa.

The objective is to minimize the *divergence* between the distributions P and Q. This yields a specific *cost function*, considered next.

4.3.1   Cost Function and Optimization

The objective of t-SNE is to find the set of coordinates z in embedded space that minimize the *Kullback-Leibler* divergence between the p_{ij} and corresponding q_{ij}. This is formalized in the following expression:

$$\min_z(C) = \min_z[KLIC(P||Q)] = \min_z \sum_{i \neq j}[p_{ij} \ln p_{ij} - p_{ij} \ln q_{ij}].$$

In essence, this boils down to trying to match pairs with a high p_{ij} (nearby points in high dimension) to pairs with a high q_{ij} (nearby points in the embedded space).

Following standard calculus, in order to minimize the cost function, its gradient is needed. This is the first partial derivative of the function C with respect to the coordinates z_i. It has the complex form:

$$\frac{\partial C}{\partial z_i} = 4 \sum_{j \neq i}(p_{ij} - q_{ij})(1 + ||z_i - z_j||^2)^{-1}(z_i - z_j).$$

The expression can be simplified somewhat by setting $U = \sum_{h \neq l}(1 + ||z_h - z_l||^2)^{-1}$, i.e., the denominator in the q_{ij} probability. As a result, since $q_{ij} = (1 + ||z_i - z_j||^2)^{-1}/U$, the term $(1 + ||z_i - z_j||^2)^{-1}$ can be replaced by $q_{ij}U$, so that the gradient becomes:

$$\frac{\partial C}{\partial z_i} = 4 \sum_{j \neq i}(p_{ij} - q_{ij})q_{ij}U(z_i - z_j).$$

The optimization involves a complex gradient search that is adjusted with a *learning rate* and a *momentum term* to speed up the process. At each iteration t, the value of z_i^t is updated as a function of its value at the previous iteration, z_i^{t-1}, the change in the function, i.e., the gradient and an adjustment that is a function of the previous change (to avoid over- and under-shooting). The contribution of the gradient is not used in full, but is multiplied by a fraction, the so-called *learning rate*, η. Since using the full gradient tends to result in inefficient changes in the result, its effect is dampened. Furthermore, there is an additional adjustment in the form of the so-called *momentum*, $\alpha(t)$, which adds a fraction of the change in the previous iteration, $(z_i^{t-1} - z_i^{t-2})$. The momentum is typically adjusted during the optimization process, hence the inclusion of the index t.

In full then, each iteration proceeds as:

$$z_i^t = z_i^{t-1} + \eta\frac{\partial C}{\partial z_i} + \alpha(t)(z_i^{t-1} - z_i^{t-2}).$$

The learning rate η and momentum $\alpha(t)$ are parameters that need to be specified. The results can vary considerably depending on how these parameters are tuned. Furthermore, the algorithm is rather finicky and may require some trial and error to find a stable solution.[1]

4.3.2   Large Data Applications (Barnes-Hut)

The most recent implementation of the t-SNE optimization process uses advanced algorithms that speed up the computations and allow the method to scale to very large data sets, as outlined in van der Maaten (2014).

The approach is based on two simplifications, one pertaining to P, the other to Q. The first strategy is to parse the distribution P and eliminate all the probabilities p_{ij} that are *too small* in some sense. The other strategy is to do something similar to the probabilities q_{ij}, in the sense that the value for several individual points that are *close* together is represented by a central point. This reduces the number of individual q_{ij} that need to be evaluated. This is referred to as the Barnes-Hut optimization (Barnes and Hut, 1986).

The t-SNE gradient is separated into two parts (in the same notation as before):

$$\frac{\partial C}{\partial z_i} = 4[\sum_{j \neq i} p_{ij}q_{ij}U(z_i - z_j) - \sum_{j \neq i} q_{ij}^2 U(z_i - z_j)],$$

where the first part (the cross-products) represents an *attractive* force and the second part (the squared probabilities) represents a *repulsive* force. The simplification of the p_{ij} targets the first part (converting many products to a value of zero), whereas the simplification of q_{ij} addresses the second part (reducing the number of expressions that need to be evaluated).

---

[1]An excellent illustration of the sensitivity of the algorithm to various tuning parameters can be found in Wattenberg et al. (2016) (https://distill.pub/2016/misread-tsne/).

4.3.2.1   Simplification of P

In order to simplify the p_{ij}, those point pairs that are farther apart than a given cut-off distance have their probability set to zero. The rationale for this is that the value of p_{ij} for those pairs that do not meet the closeness criterion is likely to be very small and hence can be ignored, similar to the logic behind Tobler's law. The distance criterion is related to the preferred *perplexity*, a rough measure of the target number of neighbors. In practice, pairs that are more than $3u$ apart, with u as the perplexity, have $p_{ij} = 0$. As a result of the choice of the distance criterion, the value for perplexity set as an option can be at most $n/3$. However, in van der Maaten and Hinton (2008), the value for perplexity is suggested to be between 5 and 50.

The parsing of p_{ij} is implemented by means of a so-called *vantage-point tree* (VP tree). In essence, this tree divides the space into those points that are closer than a given distance to a reference point (the *vantage point*) and those that are farther away. By applying this recursively, the data get partitioned into smaller and smaller entities such that neighbors in the tree are likely to be neighbors in space (see Yianilos, 1993, for technical details). The VP tree is used to carry out a nearest neighbor search so that all j that are not in the nearest neighbor set for a given i result in $p_{ij} = 0$. Importantly, the p_{ij} do not change during the optimization, since they pertain to the high-dimensional space, so this calculation only needs to be done once.

4.3.2.2   Simplification of Q

The simplification of Q relies on an efficient organization of the solution points as a quadtree, together with a replacement of points that are close together (in the same box in the quadtree) by a single center point.

At each iteration, a quadtree is created for the current layout, based on the current solution for the coordinates z.[2] The tree is traversed (depth-first) and at each node a decision is made whether the points in that cell can be effectively represented by their center. The logic behind this is that the distance from i to all the points j in the cell will be very similar, so rather than computing each individual $q_{ij}^2 U(z_i - z_j)$ for all the j in the cell, they are replaced by $n_c q_{ic}^2 (1 + ||z_i - z_c||^2)^{-1}(z_i - z_c)$, where n_c is the number of points in the cell and z_c are the coordinates of a central (representative) point. The denser the points are in a given cell, the greater the computational gain.[3]

A critical decision is the selection of those cells for which the simplification is appropriate. This is set by the θ parameter to the optimization program, such that:

$$\frac{r_c}{||z_i - z_c||} < \theta,$$

where r_c is the length of the diagonal of the cell square. In other words, the cell is summarized when the diagonal of the cell is less than θ times the distance between a point i and the center of the cell. With θ set to zero, there is no simplification going on, and all the q_{ij} are computed at each iteration.

This second simplification only makes a difference for truly large data sets.

---

[2] A quadtree is a tree data structure that recursively partitions the space into squares as long as there are points in the resulting space. For example, the bounding box of the points is first divided into four squares. If one of those is empty, that becomes an endpoint in the tree (a leaf). For those squares with points in them, the process is repeated, again stopping whenever a square is empty. This structure allows for very fast searching of the tree. In three dimensions, the counterpart (using cubes) is called an octree.

[3] For further technical details, see van der Maaten (2014).

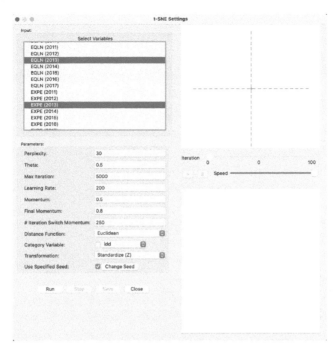

Figure 4.2: t-SNE Settings

4.4 Implementation

The t-SNE functionality is invoked from the drop-down list created by the toolbar **Clusters** icon (Figure 4.1), as the third item (more precisely, the third item in the dimension reduction category). Alternatively, from the main menu, **Clusters > t-SNE** gets the process started.

To illustrate these methods, the same ten bank indicators are used as in the previous chapter (for a full list, see Section 3.2.2).

The main dialog is the **t-SNE Settings** panel, shown in Figure 4.2. The left-hand side contains the list of available variables as well as several parameters that can be specified to fine-tune the algorithm. Both **Euclidean** and **Manhattan** distance metrics are available for the **Distance Function** (to compute the probability functions), with **Euclidean** as the default. As before, the **Standardize (Z)** is the default **Transformation**.

On the right-hand side of the interface are two major panels. The top consists of a space that is used to track the changes in the two-dimensional scatter plot as the iterations proceed. The bottom panel gives further quantitative information on the evolution of the quality of the result, reported for every 50 iterations. This lists a value for the **error**, i.e., the cost function evaluated at that stage. The table provides an indication of how the process is converging to a stable state.

In-between the two result panes on the right is a small control panel to adjust the speed of the animation, as well as allowing to pause and resume as the iterations proceed (see Section 4.4.1).

With all the parameters left to their default settings, the **Run** button populates the panels on the right-hand side. Figure 4.3 shows the result after the default 5000 iterations, with

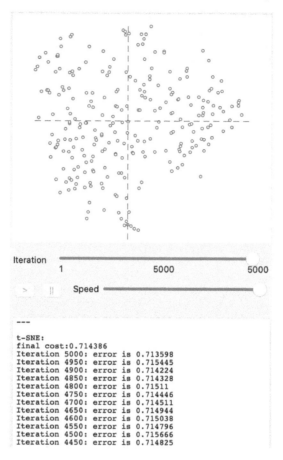

Figure 4.3: t-SNE after full iteration – default settings

a **final cost** of 0.714386. The top panel contains the familiar scatter plot of the t-SNE coordinates in two dimensions. The bottom pane (only partly shown) lists the **final cost** as well as the **error** for every 50 iterations. It shows that the error has been hovering around a similar value for (at least) the last 500 iterations.

In the implementation in `GeoDa`, the final cost value listed (0.714386) does not exactly equal the result for the error at the last iteration (0.713598). This is due to the use of the Barnes-Hut approximation (Section 4.3.2.2).[4]

With the final result in hand, selecting the **Save** button brings up the familiar dialog to specify the variable names to be saved to the data table (defaults are **V1** and **V2**, but here set to **V1tsne** and **V2tsne**). Once these coordinate variables are specified, the final scatter plot is generated in a separate window, as illustrated in Figure 4.4.

At the bottom of the graph is a listing of the variables used in the analysis, as well as the **final cost** (0.714), the **rank correlation** (0.640) and the number of **iterations** (the

---

[4]As detailed in Section 4.3.2.2, when **Theta** is not zero, several values of q_{ij} are replaced by the center of the corresponding quadtree block. The **final cost** reported is computed with the values for all $i - j$ pairs included in the calculation, and does not use the simplified q_{ij}, whereas the **error** reported for each iteration uses the simplified version.

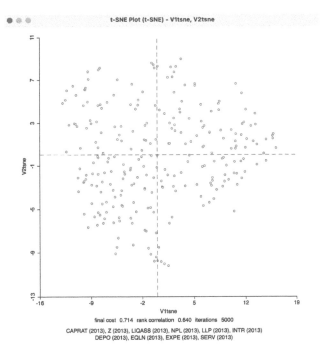

Figure 4.4: t-SNE 2D Coordinates

default 5000). Since the t-SNE optimization algorithm is not based on a specific stopping criterion, unless the process is somehow paused (see Section 4.4.1), the number of iterations will always correspond to what is specified in the options panel. However, this does not mean that it necessarily has resulted in a stable optimum. In practice, one therefore needs to pay close attention to the point pattern of the iterations as they proceed.

The **final cost** is not comparable to the objective function for MDS, since it is not based on a stress function. On the other hand, the rank correlation is a rough global measure of the correspondence between the relative distances and remains a valid measure. However, it should be kept in mind that t-SNE is geared to optimize *local* distances, so a global performance measure is not necessarily the best measure of performance.

4.4.0.1 Inspecting the iterations

The bottom panel allows for an inspection of the results for the **error** at each iteration. A more visual representation of the change in point configurations can be found by manipulating the **Iteration** button in the top panel. As the button is moved to the left, the change in configuration is illustrated in the graph as the optimization process proceeds.

This is particularly illuminating in the beginning stages of the algorithm, where substantial jumps can be observed. For example, in Figure 4.5, the **Iteration** is shown for two early results, one at iteration 250 (error 54.7166) and one shortly afterward, at iteration 300 (error 0.852884). This is the point in the solution process where initial very chaotic solutions (jumping from one side of the graph to the other) begin to stabilize to a more circular cloud shape.

This process continues until a certain degree of stability is obtained, after which the changes become marginal. For example, in Figure 4.6, the results at 2500 iterations (error 0.71491) and at 4000 iterations (error 0.715028) are shown, which portray a very similar pattern with

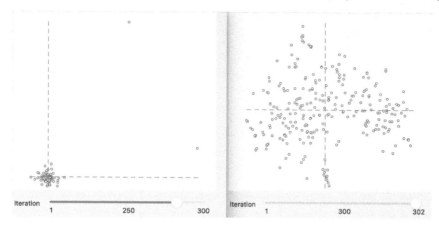

Figure 4.5: Inspecting interations of t-SNE algorithm: 250 and 300

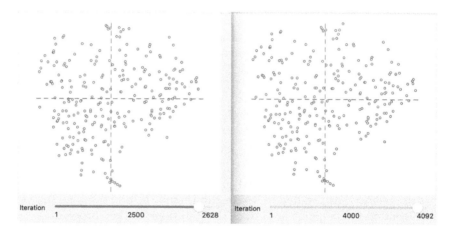

Figure 4.6: Inspecting interations of t-SNE algorithm: 1000 and 3000

almost identical error measures, near impossible to distinguish from the final solution in Figure 4.3.

4.4.1 Animation

As mentioned, with the default settings, the optimization runs its full course for the number of iterations specified in the options. The **Speed** button below the graph on the right-hand side allows for the process to be slowed down for visualization purposes, by moving the button to the left. This provides a very detailed view of the way the optimization proceeds, gets trapped in local optima and then manages to move out of them.

In addition, the pause button (||) can temporarily halt the process. The resume button (>) continues the optimization. The iteration count is given below the graph and in the bottom panel the progression of the cost minimization can be followed.

In this implementation, the pause button is different from the **Stop** button that is situated below the options in the left-hand panel. The latter ends the iterations and freezes the result, without the option to resume. However once frozen, one can move back through the previous sequence of iterations by moving the **Iteration** button to the left.

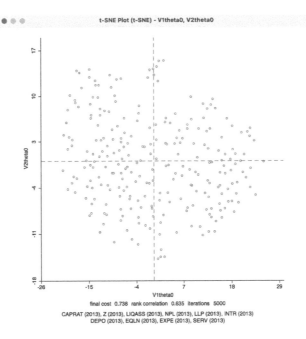

Figure 4.7: t-SNE 2D Coordinates for theta = 0

4.4.2 Tuning the Optimization

The number of options to tune the optimization algorithm may seem a bit overwhelming. However, in many instances, the default settings will be fine, although some experimenting is always a good idea.

To illustrate the effect of some of the options, changes in three central parameters are illustrated in turn: **Theta**, **Perplexity** and the **Iteration Switch** for the **Momentum**.

4.4.2.1 Theta

The Barnes-Hut algorithm implemented for t-SNE sets the **Theta** (θ) parameter to 0.5 by default. This is the criterion that determines the extent of simplification for the q_{ij} measures in the gradient equation. The parameter is primarily intended to allow the algorithm to scale up to large data sets.

In smaller data sets, like in the example considered here, this may be set to zero, which can speed up the solution since the actual distances are used, rather than a proxy. The result is shown in Figure 4.7. The **final cost** is 0.737581 with a rank correlation of 0.635, both slightly worse than the default. In this case, convergence is reached after fewer than 1000 iterations, somewhat faster than for the default case. Also, in this instance, there is no difference between the **error** at the last iteration and the **final cost**, since no Barnes-Hut simplification is carried out.

4.4.2.2 Perplexity

The second parameter to adjust is the **Perplexity** (with **Theta** set back to its default value). This is a rough proxy for the number of neighbors and is by default set to a value of 30. The results for a value of 50 (the highest suggested by van der Maaten and Hinton, 2008) are visualized in Figure 4.8. Compared to the previous results, the cost is much lower

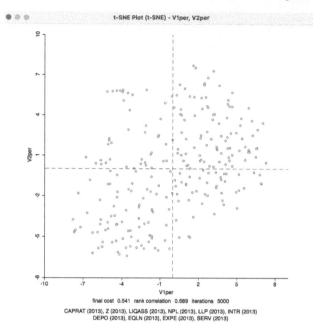

Figure 4.8: t-SNE 2D Coordinates for perplexity=50

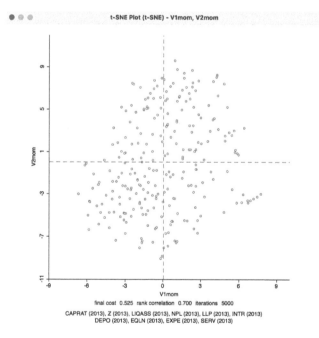

Figure 4.9: t-SNE 2D Coordinates for perplexity=50, momentum=100

at 0.540604, and a higher rank correlation of 0.689. Convergence is reached after about 1600 iterations.

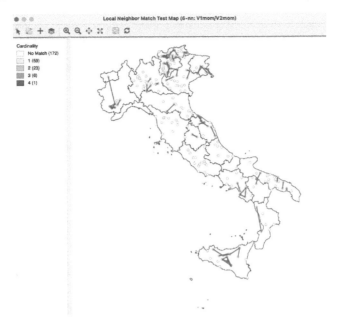

Figure 4.10: Nearest Neighbor Match Test for t-SNE

4.4.2.3 Iteration momentum switch

The final adjustment is to the **Iteration Switch** for the **Momentum** parameter, which decides at which point in the iterations the **Momentum** moves from 0.5 to 0.8. The default value is 250. With **Perplexity** left at 50, a change of the **Iteration Switch** to 100 obtains less chaotic solutions much sooner than in the previous cases. The switch happens between 100 and 150 iterations. The end result, shown in Figure 4.9, is the best achieved so far: the **final cost** is 0.525096 with a rank correlation of 0.700.

Further experimentation with the various parameters can provide deep insight into the dynamics of this complex algorithm, although the specifics will vary from case to case. Also, as argued in van der Maaten (2014), the default settings are an excellent starting point.

4.4.3 Interpretation and Spatialization

All the visualization options reviewed for classic MDS in Chapter 3 can be applied to the t-SNE solutions, in the same way as described in Sections 3.4 and 3.5. This includes the incorporation of a categorical variable that is visualized in the coordinate plot (see Section 3.4.2). In addition, for t-SNE, those categories are also shown during the iterations and animations in the top right panel of Figure 4.2. Otherwise, everything operates as discussed previously.

One aspect that remains to be investigated is the extent to which the t-SNE approach yields solutions that provide a match with geographic neighbors. This would be reflected in a nearest neighbor match test based on t-SNE nearest neighbors.

4.4.3.1 Nearest neighbor match test

The nearest neighbor match test for the coordinates of the latest t-SNE results, **V1mom** and **V2mom** is illustrated by the unique values cardinality map in Figure 4.10. The k-nearest neighbors with k=6 are matched to the geographical k-nearest neighbors. Compared to the MDS results in Figure 3.17 and the standard results in Figure 3.18, the largest number of common neighbors is four, obtained for one observation (the Cassa Raiffeisen in Latsch in the Trentino-Alto Adige Region, observation 67). The associated probability is 0.000001, suggesting very strong evidence of a multivariate local cluster. In addition, there are 6 observations with three common neighbors (p=0.000133), and 23 with two common neighbors (p=0.006275). As shown in Section 3.5.3, the result with one common neighbor cannot be deemed to be significant (p=0.125506).

The cluster analysis based on the t-SNE computations yields the most significant clusters in this example. This is somewhat to be expected, since the method stresses *local* similarities between the higher and lower embedded dimension. Many identified locations match the standard cluster results, but several are also in different locations. The involved trade-offs remain to be further investigated.

4.5 Comparing Distance Preserving Methods

Traditionally, the *fit* of a distance-preserving dimension reduction method is assessed by means of the objective function used in the algorithm, such as the stress function for MDS or cost for t-SNE. However, such measures are not comparable across methods. Another commonly used metric, the rank correlation between the original inter-observation distances and the corresponding distances in the embedded space is comparable. However, this indicator may be less appropriate for t-SNE, since that method optimizes the match between close observations. In contrast, rank correlation is a *global* correlation coefficient.

An alternative approach can be based on the concept of *common coverage percentage*, introduced in Section 3.5.2.2 as a way to measure the overlap between geographical k-nearest neighbors and k-nearest neighbors in the embedded space. It was also used to assess the overlap between different MDS solutions. The common coverage percentage is a simple ratio of the density obtained by the intersection of two k-nearest neighbor graphs to the maximum overlap possible. In the case of k-nearest neighbors, the maximum is k/n, the property listed as **% nonzero** in the **Weights Manager**.

In addition to comparing different embedded solutions, this idea can also be applied to obtain an alternative overall measure of the fit of the embedded solution, relative to the k-nearest neighbors in the multivariate attribute space. The maximum coverage is the same, since it is based on the same k for k-nearest neighbors. The only difference with the spatial case is that the reference weights are based on the neighbor structure in multi-attribute variable space. The ratio is then computed of the density of the intersection with the weights for the embedded solution to this maximum.

4.5.1 Comparing t-SNE Options

As was the case for different MDS solutions, it is not straightforward to visually compare the scatter plots that result from manipulating the parameters of the t-SNE algorithm. As before, a solution is to measure the degree of overlap between k-nearest neighbor graphs computed from the scatter plots (i.e., the scatter plot coordinates form the locational information).

For example, comparing the default t-SNE result in Figure 4.4 to the outcome from tuning the momentum and perplexity parameters in Figure 4.9 yields 1.43% nonzero elements in

	k=1	k=2	k=4	k=6	k=8	rank corr
MDS	7.83	11.75	15.66	20.45	23.49	0.80
SMACOF	10.44	13.05	18.27	24.80	28.38	0.91
t-SNE	54.81	53.51	52.85	51.77	50.90	0.64

Figure 4.11: Common Coverage Percentage measures of fit for different methods

the corresponding k-nearest neighbor graph intersection for k=6. This results in a common coverage percentage of 62.2%. This value is larger than the 46.5% found between classic metric MDS and SMACOF in Section 3.5.2.2.

4.5.2 Local Fit with Common Coverage Percentage

As mentioned, the common coverage percentage logic can be used to compute a *local* measure of goodness of fit between the embedded space nearest neighbors and the multi-attribute nearest neighbors.

Figure 4.11 shows the results for classic metric MDS (Figure 3.3), SMACOF (using Euclidean distances, Figure 3.8) and default t-SNE (Figure 4.4). The common coverage percentage is computed for values of k equal to 1, 2, 4, 6 and 8, corresponding with increasingly less local neighborhoods. In addition, for the sake of completeness, the rank correlation is listed as well.

As we had already seen previously, SMACOF does best on the rank correlation measure, with t-SNE yielding the worst results. In contrast, t-SNE does much better on the measures of local fit. The focus on the *local* is illustrated by a slightly decreasing coverage percentage, from 54.81% for the nearest neighbor (k=1) to 50.90% for k=8. These percentages indicate more than 50% match between the nearest neighbors. The classic MDS methods do much worse, with SMACOF consistently outperforming metric MDS, but achieving clearly inferior measures of local fit relative to t-SNE. As is to be expected, the common coverage percentage improves with the number of neighbors, almost tripling for SMACOF, from 10.44% for k=1 to 28.38% for k=8.

These findings clearly illustrate the strong focus on local matches inherent in the t-SNE algorithm.

Part II

Classic Clustering

5

Hierarchical Clustering Methods

At this point, the focus shifts from reducing the dimensionality of the variables to reducing the dimensionality of observations, through the *clustering* of *similar* observations into a smaller number of groups. The chapters in this Part deal with classic clustering methods, which are inherently nonspatial. However, as before, a strong spatial perspective is maintained by linking the solutions to a geographical representation. Part III covers clustering methods where spatial contiguity of cluster members is enforced as a constraint.

In general terms, clustering methods group n observations into k *clusters*, such that both the *intra-cluster similarity and the between-cluster dissimilarity are maximized*. Equivalently, one can think of it as minimizing intra-cluster dissimilarity and between-cluster similarity, the complement of the previous objective.

The goal of clustering methods is thus to achieve compact groups of *similar observations* in multi-attribute space that are separated as much as possible from the other groups.[1]

There are a very large number of clustering techniques and algorithms. These methods are standard tools of so-called *unsupervised learning*. They constitute a core element in any machine learning toolbox. Classic texts include Hartigan (1975), Jain and Dubes (1988), Kaufman and Rousseeuw (2005) and Everitt et al. (2011). A fairly recent historical overview of method development can be found in Jain (2010). In addition, excellent treatments of some of the more technical aspects are contained in Chapter 14 of Hastie et al. (2009), Chapters 10 and 11 of Han et al. (2012) and Chapter 10 of James et al. (2013), among others.

Clustering methods can be organized along a number of different dimensions. A common distinction is between hierarchical methods, partitioning methods, density-based methods and grid-based methods (see, e.g., Han et al., 2012, pp. 448–450). In addition, there are model-based approaches developed in the statistical literature, such as Gaussian mixture models (GMM) and Bayesian clusters.

In this and the next three chapters, the essence behind two most common approaches is covered, namely *hierarchical* methods and *partitioning* methods. In addition, a more recent approach based on *spectral* decomposition is treated as well. Model-based techniques are not considered, since they are less compatible with the exploratory mindset taken in these Volumes. More precisely, they require a formal probabilistic model.

In addition, the discussion is limited to exact clustering methods. Therefore fuzzy clustering techniques are not considered, i.e., methods that yield solutions where an observation may belong to more than one cluster. In exact clustering, the clusters are both *exhaustive* and *exclusive*. This means that every observation must belong to one cluster, and only one cluster.

---

[1]While dimension reduction and clustering are treated here separately, so-called *biclustering* techniques group both variables and observations simultaneously. While old, going back to an article by Hartigan (1972), these techniques have gained a lot of interest more recently in the field of gene expression analysis. For overviews, see, e.g., Tanay et al. (2004) and Padilha and Campello (2017).

DOI: 10.1201/9781032713175-5

The topic of this chapter is *Hierarchical* clustering. *Partitioning* clustering is covered in the next chapter, more advanced techniques in the third and the fourth chapter is devoted to *Spectral* clustering.

Hierarchical clustering methods build up the clusters step by step. The size of the cluster is not set a priori (in contrast to what is the case for partitioning methods), but is determined from the results at each step that are visually represented in a tree structure, the so-called *dendrogram*.

The successive steps can be approached in a top-down fashion or in a bottom-up fashion. The latter is referred to as *agglomerative* hierarchical clustering, the former as *divisive* clustering. In this chapter, only *agglomerative* methods are treated. A form of divisive clustering is implemented in some of the spatially constrained clustering techniques discussed in Chapter 10.

The chapter begins with a discussion of the concept of dissimilarity, which is essential to all clustering procedures. Next, agglomerative clustering is introduced and the importance of the so-called *linkage function* is outlined. The visual expression of clusters is obtained by means of a *dendrogram*. Four important linkage functions are discussed: single linkage, complete linkage, average linkage and Ward's method.

The methods are illustrated by means of the *Chicago Community Areas* sample data set that contains several socio-economic indicators for 2020. This particular example was chosen because the 77 observations allow for an easy visual analysis of the dendrogram. With larger data sets, such as the *Chicago SHOH* sample data set (n = 791) considered in the other cluster chapters, it becomes difficult to visualize the details. This aspect of the dendrogram is one of the drawbacks of hierarchical clustering applications for large(r) data sets.

5.1   Topics Covered

- Understand the principles behind agglomerative clustering
- Distinguish between different linkage functions
- Understand the implications of using different linkage functions
- Interpret a dendrogram for hierarchical clustering
- Interpret cluster characteristics
- Visualize the clusters in a cluster map
- Carry out sensitivity analysis

GeoDa Functions

- Clusters > Hierarchical
    - select cut point
    - select linkage function
    - select standardization option
    - select distance metric
    - cluster characteristics
    - mapping the clusters
    - saving the cluster classification

Toolbar Icons

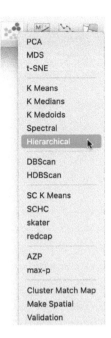

Figure 5.1: Clusters > K Means | K Medians | K Medoids | Spectral | Hierarchical

5.2 Dissimilarity

The overall objective in any clustering exercise is to end up with a grouping that minimizes the dissimilarity *within* each cluster.

Mathematically, the point of departure is an overall loss function that consists of summing the *distances* between all the pairs of observations:

$$T = (1/2) \sum_{i=1}^{n} \sum_{j=1}^{n} d_{ij},$$

where d_{ij} is some measure of dissimilarity, such as the Euclidean distance between the values at observations i and j.[2]

A cluster assignment can be symbolized by an *encoder* C, such that each observation i is assigned to a cluster, $C(i) = h$, where the cluster indicator h is an element from the set $\{1, \ldots, k\}$. The cluster labels themselves (h) are meaningless, and could just as well be letters or other distinct symbols. In agglomerative clustering, the assignment is bottom-up, with k starting at n (each observation is its own cluster) and eventually ending up at 1 (all observations are in a single cluster).

---

[2]Since each pair is counted twice, the total sum is divided by 2. While this seems arbitrary at this point, it helps in some calculations. For hierarchical clustering methods, the distinction is immaterial.

At each stage, the loss function can be evaluated. In any given cluster h, the distances from one of its members ($i \in h$) to all other observations can be separated out between those that belong to the cluster ($j \in h$) and those that do not ($j \notin h$):

$$T_i = (1/2)[\sum_{j \in h} d_{ij} + \sum_{j \notin h} d_{ij}].$$

This can be generalized to all the elements of cluster h by summing over i:

$$T_{i \in h} = (1/2)[\sum_{i \in h} \sum_{j \in h} d_{ij} + \sum_{i \in h} \sum_{j \notin h} d_{ij}].$$

More generally, for all the cluster sizes up to k, i.e., for all possible pairs, the sum is over h:

$$T = (1/2)(\sum_{h=1}^{k} [\sum_{i \in h} \sum_{j \in h} d_{ij} + \sum_{i \in h} \sum_{j \notin h} d_{ij}]) = W + B,$$

where the first term (W) is referred to as the *within* dissimilarity and the second (B) as the *between* dissimilarity.

In other words, the total dissimilarity decomposes into one part due to what happens within each cluster and another part that pertains to the between cluster dissimilarities. W and B are complementary, since T is fixed. Therefore, the lower W, the higher will be B and vice versa.

In hierarchical clustering, the loss function is not optimized per se, but rather it is evaluated for each k that results from a *cut* in the dendrogram (see Section 5.3.2).

5.3   Agglomerative Clustering

An agglomerative clustering algorithm starts with each observation serving as its own cluster, i.e., beginning with n clusters of size 1. Next, the algorithm moves through a sequence of steps, where each time the number of clusters is decreased by one, either by creating a new cluster by joining two individual observations, by assigning an observation to an existing cluster or by merging two clusters. Such algorithms are sometimes referred to as SAHN, which stands for sequential, agglomerative, hierarchic and nonoverlapping (Müllner, 2011).

The sequence of merging observations into clusters is graphically represented by means of a tree structure, the so-called *dendrogram* (see Section 5.3.2). At the bottom of the tree, the individual observations constitute the *leaves*, whereas the *root* of the tree is the single cluster that consists of all observations.

The smallest within-group sum of squares is obtained in the initial stage, when each observation is its own cluster. As a result, the within sum of squares is zero and the between sum of squares is at its maximum, which also equals the total sum of squares. As soon as two observations are grouped, the within sum of squares increases. Hence, each time a new merger is carried out, the overall objective of minimizing the within sum of squares deteriorates. At the final stage, when all observations are joined into a single cluster, the total within sum of squares now also equals the total sum of squares, since there is no between sum of squares (the two are complementary).

In other words, in an agglomerative hierarchical clustering procedure, the objective function gets worse at each step. It is therefore not optimized as such, but instead can be evaluated at each step.

One distinguishing characteristic of hierarchical clustering is that once an observation is grouped with other observations, it cannot be disassociated from them in a later step. This precludes *swapping* of observations between clusters, which is a characteristic of some of the partitioning methods. This property of getting *trapped* into a cluster (i.e., into a branch of the dendrogram) can be limiting in some contexts.

5.3.1 Linkage and Updating Formula

A key aspect in the agglomerative process is how to define the distance between clusters, or between a single observation and a cluster. This is referred to as the *linkage*. There are at least seven different concepts of linkage, but here only the four most common ones are considered: single linkage, complete linkage, average linkage and Ward's method.

A second important concept is how the distances between other points (or clusters) and a newly merged cluster are computed, the so-called *updating formula*. With some clever algebra, it can be shown that these calculations can be based on the dissimilarity matrix from the previous step. The update thus does not require going back to the original $n \times n$ dissimilarity matrix.[3] Moreover, at each step, the dimension of the relevant dissimilarity matrix decreases by one, which allows for very memory-efficient algorithms.

Each linkage type and its associated updating formula is briefly considered in turn.

5.3.1.1 Single linkage

For *single linkage*, the relevant dissimilarity is between the two points in each cluster that are closest together. More precisely, the dissimilarity between clusters A and B is:

$$d_{AB} = \min_{i \in A, j \in B} d_{ij},$$

The updating formula yields the dissimilarity between a point (or cluster) P and a cluster C that was obtained by merging A and B. It is the smallest of the dissimilarities between P and A and P and B:

$$d_{PC} = \min(d_{PA}, d_{PB}).$$

The minimum condition can also be obtained as the result of an algebraic expression, which yields the updating formula as:

$$d_{PC} = (1/2)(d_{PA} + d_{PB}) - (1/2)|d_{PA} - d_{PB}|,$$

in the same notation as before.[4]

The updating formula only affects the row/column in the dissimilarity matrix that pertains to the newly merged cluster. The other elements of the dissimilarity matrix remain unchanged.

Single linkage clusters tend to result in a few clusters consisting of long drawn-out chains of observations, in combination with several singletons (observations that form their own cluster). This is due to the fact that disparate clusters may be joined when they have two close points, but otherwise are far apart. Single linkage is sometimes used to detect outliers, i.e., observations that remain singletons and thus are far apart from all others.

---

[3]Detailed proofs for all the properties are contained in Chapter 5 of Kaufman and Rousseeuw (2005).

[4]To see that this holds, consider the situation when $d_{PA} < d_{PB}$, i.e., A is the nearest neighbor to P. As a result, the absolute value of $d_{PA} - d_{PB}$ is $d_{PB} - d_{PA}$. Then the expression becomes $(1/2)d_{PA} + (1/2)d_{PB} - (1/2)d_{PB} + (1/2)d_{PA} = d_{PA}$, the desired result.

5.3.1.2   Complete linkage

Complete linkage is the opposite of single linkage in that the dissimilarity between two clusters is defined as the farthest neighbors, i.e., the pair of points, one from each cluster, that are separated by the greatest dissimilarity. For the dissimilarity between clusters A and B, this boils down to:

$$d_{AB} = \max_{i \in A, j \in B} d_{ij}.$$

The updating formula is the opposite of the one for single linkage. The dissimilarity between a point (or cluster) P and a cluster C that was obtained by merging A and B is the largest of the dissimilarities between P and A and P and B:

$$d_{PC} = \max(d_{PA}, d_{PB}).$$

The algebraic counterpart of the updating formula is:

$$d_{PC} = (1/2)(d_{PA} + d_{PB}) + (1/2)|d_{PA} - d_{PB}|,$$

using the same logic as in the single linkage case.

In contrast to single linkage, complete linkage tends to result in a large number of well-balanced compact clusters. Instead of merging fairly disparate clusters that have (only) two close points, it can have the opposite effect of keeping similar observations in separate clusters.

5.3.1.3   Average linkage

In *average linkage*, the dissimilarity between two clusters is the average of all pairwise dissimilarities between observations i in cluster A and j in cluster B. There are $n_A.n_B$ such pairs (only counting each pair once), with n_A and n_B as the number of observations in each cluster. Consequently, the dissimilarity between A and B is (without double counting pairs in the numerator):

$$d_{AB} = \frac{\sum_{i \in A} \sum_{j \in B} d_{ij}}{n_A.n_B}.$$

In the special case when two single observations are merged, d_{AB} is simply the dissimilarity between the two, since $n_A = n_B = 1$ and thus the denominator in the expression is 1.

The updating formula to compute the dissimilarity between a point (or cluster) P and the new cluster C formed by merging A and B is the weighted average of the dissimilarities d_{PA} and d_{PB}:

$$d_{PC} = \frac{n_A}{n_A + n_B} d_{PA} + \frac{n_B}{n_A + n_B} d_{PB}.$$

As before, the other distances are not affected.[5]

Average linkage can be viewed as a compromise between the nearest neighbor logic of single linkage and the furthest neighbor logic of complete linkage.

5.3.1.4   Ward's method

The three linkage methods discussed so far only make use of a dissimilarity matrix. How this matrix is obtained does not matter. As a result, dissimilarity may be defined using Euclidean or Manhattan distance, dissimilarity among categories or even directly from interview or survey data.

---

[5]By convention, the diagonal dissimilarity for the newly merged cluster is set to zero.

In contrast, the method developed by Ward (1963) is based on a sum of squared errors rationale that only works for Euclidean distance between observations. In addition, the sum of squared errors requires the consideration of the so-called *centroid* of each cluster, i.e., the mean vector of the observations belonging to the cluster. Therefore, the input into Ward's method is not a dissimilarity matrix, but a $n \times p$ matrix X of n observations on p variables (as before, this is typically standardized in some fashion).

Ward's method is based on the objective of minimizing the deterioration in the overall within sum of squares. The latter is the sum of squared deviations between the observations in a cluster and the centroid (mean):

$$WSS = \sum_{i \in C} (x_i - \bar{x}_C)^2,$$

with $\bar{x}_C$ as the centroid of cluster C.

Since any merger of two existing clusters (including the merger of individual observations) results in a worsening of the overall WSS, Ward's method is designed to minimize this deterioration. More specifically, it is designed to minimize the difference between the new (larger) WSS in the merged cluster and the sum of the WSS of the components that were merged. This turns out to boil down to minimizing the distance between cluster centers.[6] Without loss of generality, it is easier to express the dissimilarity in terms of the square of the Euclidean distance:

$$d_{AB}^2 = \frac{2n_A n_B}{n_A + n_B} ||\bar{x}_A - \bar{x}_B||^2,$$

where $||\bar{x}_A - \bar{x}_B||$ is the Euclidean distance between the two cluster centers (squared in the distance squared expression).[7]

The update equation to compute the (squared) distance from an observation (or cluster) P to a new cluster C obtained from the merger of A and B is more complex than for the other linkage options:

$$d_{PC}^2 = \frac{n_A + n_P}{n_C + n_P} d_{PA}^2 + \frac{n_B + n_P}{n_C + n_P} d_{PB}^2 - \frac{n_P}{n_C + n_P} d_{AB}^2,$$

in the same notation as before. However, it can still readily be obtained from the information contained in the dissimilarity matrix from the previous step, and it does not involve the actual computation of centroids.

To see why this is the case, consider the usual first step when two single observations are merged. The distance squared between them is simply the Euclidean distance squared between their values, not involving any centroids. The updated squared distances between other points and the two merged points only involve the point-to-point squared distances d_{PA}^2, d_{PB}^2 and d_{AB}^2, no centroids. From then on, any update uses the results from the previous distance matrix in the update equation.

5.3.1.5 Illustration – single linkage

To illustrate the logic of agglomerative hierarchical clustering algorithms, the *single linkage* approach is applied to a toy example consisting of the coordinates of seven points, shown

---

[6]See Kaufman and Rousseeuw (2005), Chapter 5, for detailed proofs.

[7]The factor 2 is included to make sure the expression works when two single observations are merged. In such an instance, their centroid is their actual value and $n_A + n_B = 2$. It does not matter in terms of the algorithm steps.

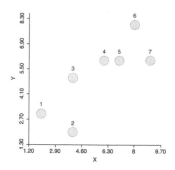

Figure 5.2: Single Linkage Hierarchical Clustering Toy Example

ID	X	Y
1	2	3
2	4	2
3	4	5
4	6	6
5	7	6
6	8	8
7	9	6

Figure 5.3: Coordinate Values

	1	2	3	4	5	6	7
1	0.00	2.24	2.83	5.00	5.83	7.81	7.62
2		0.00	3.00	4.47	5.00	7.21	6.40
3			0.00	2.24	3.16	5.00	5.10
4				0.00	1.00	2.83	3.00
5					0.00	2.24	2.00
6						0.00	2.24
7							0.00

Figure 5.4: Single Linkage – Step 1

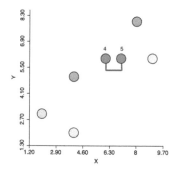

Figure 5.5: Smallest nearest neighbors

in Figure 5.2. The corresponding X, Y values are listed in Figure 5.3. The point IDs are ordered with increasing values of X first, then Y, starting with observation 1 in the lower left corner.

The basis for any agglomerative clustering method is a $n \times n$ symmetric dissimilarity matrix. Except for Ward's method, this is the only information needed. The dissimilarity matrix derived from the Euclidean distances between the points is shown in Figure 5.4. Since the matrix is symmetric, only the upper diagonal elements are listed.

The first step in the algorithm is to identify the pair of observations that have the smallest nearest neighbor distance. In the distance matrix, this is the row-column combination with

	1	2	3	4,5	6	7
1	0.00	2.24	2.83	5.00	7.81	7.62
2		0.00	3.00	4.47	7.21	6.40
3			0.00	2.24	5.00	5.10
4,5				0.00	2.83	2.00
6					0.00	2.24
7						0.00

Figure 5.6: Single Linkage – Step 2

	1	2	3	4,5,7	6
1	0.00	2.24	2.83	5.00	7.81
2		0.00	3.00	4.47	7.21
3			0.00	2.24	5.00
4,5,7				0.00	2.24
6					0.00

Figure 5.7: Single Linkage – Step 3

	1,2	3	4,5,7	6
1,2	0.00	2.83	4.47	7.21
3		0.00	2.24	5.00
4,5,7			0.00	2.24
6				0.00

Figure 5.8: Single Linkage – Step 4

the smallest entry. In Figure 5.4 this is readily identified as the pair 4-5 ($d_{4,5} = 1.0$). This pair therefore forms the first cluster, connected by a link in Figure 5.5. The two points in the cluster are highlighted in dark blue. The five other observations remain in their initial separate cluster. In other words, at this stage, there are six clusters, one less than the number of observations.

The dissimilarity matrix is updated using the smallest dissimilarity between each observation and either observation 4 or observation 5. This yields the updated row and column entries for the combined unit 4,5. More precisely, the dissimilarity used between the cluster and the other observations varies depending on whether observation 4 or 5 is closest to the other observations. For example, in Figure 5.6, the dissimilarity between 4,5 and 1 is given as 5.0, which is the smallest of 1-4 (5.0) and 1-5 (5.83). The dissimilarities between the pairs of observations that do not involve 4,5 are not affected.

The other entries for 4,5 are updated in the same way, and again the smallest dissimilarity is located in the matrix. This time, it is a dissimilarity of 2.0 between 4,5 and 7 (more precisely, between the closest pair 5 and 7). Consequently, observation 7 is added to the 4,5 cluster.

The dissimilarities between 4,5,7 and the other points are updated in Figure 5.7. However, now there is a problem. There is a three-way tie in terms of the smallest value: 1-2, 4,5,7-3 and 4,5,7-6 all have a dissimilarity of 2.24, but only one can be picked to update the clusters. Ties are typically handled by choosing one grouping at random. In the example, the pair 1-2 is selected, which is how the tie is broken by the algorithm used in GeoDa.[8]

With the distances updated, not unsurprisingly, 2.24 is again found as the shortest dissimilarity, tied for two pairs (in Figure 5.8). This time the algorithm adds 3 to the existing cluster 4,5,7.

---

[8]An implementation of the **fastcluster** algorithm.

	1,2	4,5,7,3	6
1,2	0.00	2.83	7.21
4,5,7,3		0.00	2.24
6			0.00

	1,2	4,5,7,3,6
1,2	0.00	2.83
4,5,7,3,6		0.00

Figure 5.9: Single Linkage – Step 5

Figure 5.10: Single Linkage – Step 6

Figure 5.11: Single linkage iterations

Finally, observation 6 is added to cluster 4,5,7,3 again for a dissimilarity of 2.24 (in Figure 5.9).

The end result is to merge the two clusters 1-2 and 4,5,7,3,6 into a single one, which ends the iterations (Figure 5.10).

In sum, the algorithm moves sequentially to identify the nearest neighbor at each step, merges the relevant observations/clusters and so decreases the number of clusters by one. The sequence of steps is illustrated in the panels of Figure 5.11, going from left to right and starting at the top.

5.3.2 Dendrogram

While the visual representation of the sequential grouping of observations in Figure 5.11 works well in this toy example, it is not practical for larger data sets.

A tree structure that visualizes the agglomerative *nesting* of observations into clusters is the so-called *dendrogram*. For each step in the process, the graph shows which observations/clusters are combined. In addition, the degree of change in the objective function achieved by each merger is visualized by a corresponding distance on the horizontal (or vertical) axis.

The implementation of the dendrogram in GeoDa is currently somewhat limited, but it accomplishes the main goal. In Figure 5.12, the dendrogram is illustrated for the single

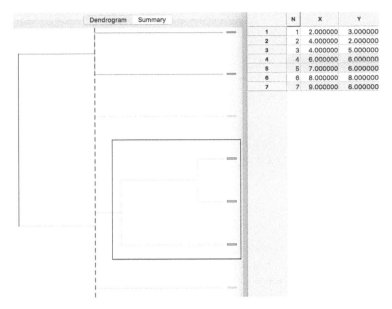

Figure 5.12: Single linkage dendrogram

linkage method in the toy example. The graph shows how the cluster starts by combining two observations (4 and 5), to which then a third (7) is added. These first two steps are contained inside the highlighted black square in the figure. The corresponding observations are selected as entries in a matching data table.

Next, following the tree structure reveals how two more observations (1 and 2) are combined into a separate cluster, and two observations (3 and 6) are added to the original cluster of 4,5 and 7. Given the three-way *tie* in the inter-group distances, the last three operations all line up (same distance from the right side) in the graph. As a result, the change in the objective function (more precisely, a deterioration) that follows from adding the points to a cluster is the same in each case.

The dashed vertical line represents a *cut* line. It corresponds with a particular value of k for which the make up of the clusters and their characteristics can be further investigated. As the cut line is moved, the members of each cluster are revealed that correspond with a different value for k.

In practice, important cluster characteristics are computed for each of the selected cut points, such as the sum of squares, the total within sum of squares, the total between sum of squares, and the ratio of the total between sum of squares to the total sum of squares (the higher ratio, the better). This will be further illustrated as part of the discussion of implementation issues in the next section.

5.4 Implementation

As mentioned in the introduction, a relatively small data set is used to illustrate the hierarchical clustering methods. The collection of socio-economic determinants of health (SDOH) for 77 Chicago Community Areas allows for a ready visualization and interpretation

Variable	Abbrev	Mean	St.Dev.	Min	Max
Per capita income	INCPERCAP	32501	19482	11857	101727
Aged < 20	Und20	24.6	6.4	8.9	44.3
Aged > 65	Ov65	13.7	4.6	4.8	26.5
Ethnic/racial minority	Minrty	72.0	26.9	16.8	99.5
Unemployment rate	Unempt	10.6	6.9	0.4	30.4
No vehicle	Noveh	12.7	8.0	1.9	34.3
No high school	Lt_high	16.0	9.6	1.7	44.7
Renters	Rentocc	45.2	15.7	9.1	76.9
Rent burden	Rntburd	25.7	8.3	11.2	49.8
Limited English proficiency	Noeng	33.4	26.7	2.2	86.5

Figure 5.13: CMAP SDOH Variables Descriptive Statistics

of the dendrogram. For larger data sets, it becomes more difficult to distinguish details in the dendrogram, although the selection of clusters by means of a *cut* line remains operational.

The variables in the *Chicago Community Areas* sample data set form a subset of the variables identified in Kolak et al. (2020) in an analysis that includes all U.S. census tracts. A subset of those census tract data will be used in the remaining chapters, for the 791 census tracts in Chicago. While not all census tract variables have a counterpart in the Chicago Community Area statistical snapshots, the match is close enough for the illustration.

Ten variables are selected:

- INCPERCAP: per capita income
- Und20: population share aged < 20
- Ov65: population share aged > 65
- Unempt: unemployment rate
- Noveh: percentage of no vehicle in the household
- Lt_high: percentage without high school education
- Rentocc: percentage renter-occupied housing
- Rntburd: rent burden
- Noeng: limited English proficiency

Except for income per capita and rent burden (ratio of 12 times the median monthly rent over median income), all values are percentages computed from the original data published by the Chicago Area Metropolitan Agency for Planning (CMAP), based on the 2014 American Community Survey data.

Simple descriptive statistics are provided in Figure 5.13. Only per capita income and rent burden show several outlying observations at the high end. The other variables have no outliers or only a single one (Und20, Ov65 and Lt_high).

5.4.1   Variable Settings Dialog

Hierarchical clustering is invoked from the drop-down list associated with the cluster toolbar icon, as the last item in the classic clustering methods subset, shown in Figure 5.1. It can also be selected from the menu as **Clusters > Hierarchical**.

This brings up the **Hierarchical Clustering Settings** dialog, which consists of two main panels. The left panel contains the interface to specify the variables and a range of parameters, shown in Figure 5.14. The right panel provides the results, either as a **Dendrogram** or as a tabular **Summary**, each invoked by a button at the top.

Figure 5.14: Hierarchical Clustering Variable Settings

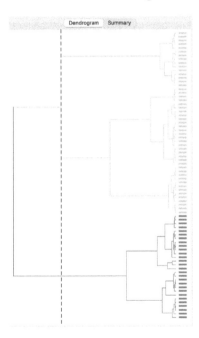

Figure 5.15: Initial Dendrogram – Ward's Method

In the example, the ten variables are selected, with all other items left to their default settings. This means that the initial number of clusters obtained (i.e., the location of the cut line) is set to **2**, the method is **Ward's-linkage** and the variables are transformed to **Standardize (Z)**. Since Ward's method only works for Euclidean distance, the **Distance Function** is not available as an option.

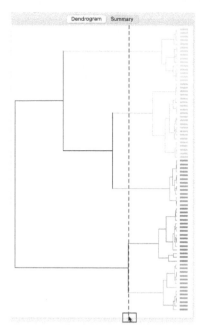

Figure 5.16: Dendrogram – Ward's Method, k=5

```
Number of clusters: 5
Transformation: Standardize (Z)
Method: Ward's-linkage
Distance function: Euclidean

Cluster centers:
   | INCPERCAP|Und20  |Ov65   |Minrty |Unemprt|Noveh  |Lt_high|Rentocc|Rntburd|Noeng
-- |----------|-------|-------|-------|-------|-------|-------|-------|-------|-------
C1 | 34837.8  |24.1731|14.6649|50.9895|5.45916|5.66421|13.6642|33.3726|19.0464|45.7101
C2 | 22405.9  |23.3887|19.4392|94.7542|16.6052|13.7278|16.409 |40.4355|28.288 |12.5214
C3 | 62399.1  |16.6054|10.7945|42.527 |4.2886 |18.0471|5.73117|54.3308|21.5517|25.2568
C4 | 19440.4  |29.9876|11.9153|96.4308|18.7494|22.9282|16.2022|62.4673|38.8783|7.23255
C5 | 18938.7  |30.2941|10.6131|89.5507|10.8771|6.68741|30.9194|41.9807|25.1344|73.026

The total sum of squares:  760
Within-cluster sum of squares:
   | Within cluster S.S.
-- |--------------------
C1 | 80.4374
C2 | 67.1242
C3 | 56.5722
C4 | 44.3585
C5 | 22.0003

The total within-cluster sum of squares:   270.493
The between-cluster sum of squares:     489.507
The ratio of between to total sum of squares:  0.644089
```

Figure 5.17: Summary – Ward's Method, k=5

5.4.2   Ward's method

Clicking on the **Run** button generates the initial dendrogram in the right-hand panel, shown in Figure 5.15. At the very right, each observation is symbolized by a small rectangle, from which the subsequent groupings are shown as a graph. Each cluster formation is represented by a vertical line relative to the horizontal axis, with groupings closer to the right achieving a smaller total within sum of squares. As the number of clusters decreases to a manageable degree, the within sum of squares increases.

The red dashed line corresponds with the cut line, set to obtain 2 clusters (the default). The two clusters are shown with different colors for the rectangles on the right (in the illustration, light blue and dark blue). Linking and brushing is implemented, so that any

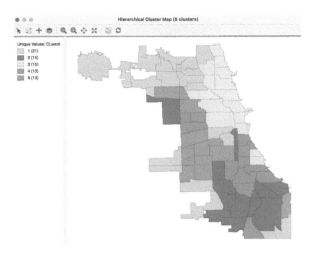

Figure 5.18: Cluster Map – Ward's Method, k=5

subset of observations can be selected among the rectangles and identified on all the open maps and graphs.

Typically, the initial solution is not of much interest. To obtain other cluster sizes, the cut line can be moved by *grabbing* it, illustrated by the highlighted pointer in Figure 5.16, for $k=5$. Alternatively, the **Number of Clusters** can also be changed in the dialog, which automatically will move the cut line to the corresponding location and adjust the cluster groupings of the individual observations.

At this point, invoking **Save/Show Map** has three direct results: the cluster classification is saved in the data table using the variable name specified in **Save Cluster in Field** (in the example, this is set to **CLward**); the **Summary** table becomes available; and a **Cluster Map** is produced as a unique values map with the cluster classifications.

As mentioned, the classification labels are meaningless. The convention used in GeoDa is to label the largest cluster as **1** with subsequent labels in decreasing order of size.

The summary table in the example is shown in Figure 5.17, with the corresponding cluster map in Figure 5.18.

For each cluster, the cluster centers are given as the average for that variable computed from the observations in the cluster.[9] This allows for a substantive interpretation, although that is not always straightforward. For example, cluster 3 (C3) achieves the highest value for income per capita, and the lowest entries for under 20, percent minority, unemployment rate and less than high school. This would suggest a higher income cluster, but it has the second highest value for absence of cares and percent renters, and the third for percent not English speaking. Clearly, the interpretation and *labeling* of the resulting clusters require some creativity and imagination.

Below the cluster centers follows **The total sum of squares**, which here is **760**, with the **Within-cluster sum of squares** listed for each cluster.[10] At the bottom of the summary are the overall statistics, including the total **within-cluster sum of squares**, at **270.493**

---

[9]Note that the summary shows the cluster centers in the original scale, whereas the clustering operation itself is typically carried out on standardized variables.

[10]The sum of squares measures are based on the standardized variables.

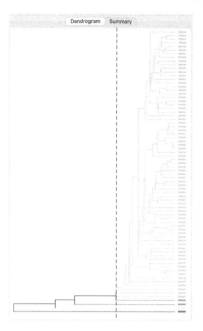

Figure 5.19: Dendrogram – Single Linkage, k=5

(the sum of the individual cluster entries), the **between-cluster sum of squares**, at **489.507** and the **ratio of between to total sum of squares**, at **0.644089**. A higher ratio indicates better cluster performance.

A useful option is to **Save** the cluster results to a text file. This is selected by right-clicking on the panel.

The cluster map depicts the location of the cluster members and the cardinality of each cluster. This is a way to *spatialize* the outcome of an otherwise nonspatial clustering method.

The largest cluster contains 21 community areas, followed by two clusters with 15 and two with 13 members. In this example, the clusters are also fairly spatially compact, with cluster 3 consisting largely of community areas in the north-east of the city, except for two community areas (Kenwood and Hyde Park) that form a small enclave surrounded by cluster 4. The latter is characterized by a high percentage minority, high unemployment and rent burden, in stark contrast with the characteristics of cluster 3.

Setting the number of clusters at five is by no means necessarily the best solution. In an actual application of hierarchical clustering, one would experiment with different cut points and evaluate the solutions.

For completeness sake, the dendrogram, summary and cluster maps are shown next for the three other linkage methods.

5.4.3 Single linkage

The linkage options are chosen from the **Method** item in the dialog. The cluster results for single linkage are typically characterized by one or a few very large clusters and several singletons (one observation per cluster). This characteristic is confirmed in the example, with the dendrogram shown in Figure 5.19, for $k=5$. The graph indicates how all but four

```
Number of clusters: 5
Transformation: Standardize (Z)
Method: Single-linkage
Distance function: Euclidean

Cluster centers:
   |INCPERCAP|Und20   |Ov65   |Minrty |Unempt |Novoh  |Lt_high|Rentooc|Rntburd|Noeng
-- |---------|--------|-------|-------|-------|-------|-------|-------|-------|-------
C1 |33320.4  |24.5918 |13.3117|70.7815|10.0111|12.042 |15.6431|44.51  |25.4866|34.049
C2 |21858    |18.5395 |24.8198|86.0494|8.20697|19.9246|37.1823|61.1907|27.2874|69.4732
C3 |11857    |44.2783 |4.7975 |98.7904|30.4174|30.3275|15.6869|76.9207|30.7093|3.09423
C4 |16519    |19.883  |26.4411|96.3659|23.622 |34.3088|21.3999|56.1983|31.2597|7.38462
C5 |19972    |13.3101 |26.5204|99.5015|20.1403|16.9492|15.4145|36.605 |29.955 |4.414

The total sum of squares:  760
Within-cluster sum of squares:
   |Within cluster S.S.
-- |-------------------
C1 |668.502
C2 |0
C3 |0
C4 |0
C5 |0

The total within-cluster sum of squares:  668.502
The between-cluster sum of squares:   91.4975
The ratio of between to total sum of squares:  0.120391
```

Figure 5.20: Summary – Single Linkage, k=5

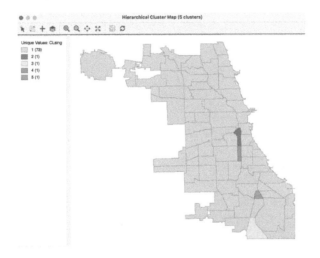

Figure 5.21: Cluster Map – Single Linkage, k=5

observations are grouped into a single cluster. This situation is not remedied by moving the cut point such that more clusters result, since almost all of the additional clusters are singletons as well.

The summary in Figure 5.20 lists a within cluster sum of squares of zero for the four singletons. The overall fit is poor, with a between to total ratio of only 0.120291.

The cluster map in Figure 5.21 highlights the four *outliers*. In other respects, this map is uninteresting, given the large size of the dominant cluster.

In practice, in most situations, single linkage will not be a good choice, unless the objective is to identify a lot of singletons and characterize these as *outliers*.

5.4.4 Complete linkage

The complete linkage method yields clusters that are similar in balance to Ward's method. For example, in Figure 5.22, the dendrogram is shown, using a cut point with five clusters. While much more balanced than single linkage, the smallest clusters are much smaller than the others.

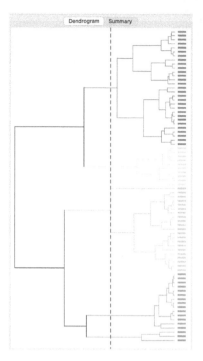

Figure 5.22: Dendrogram – Complete Linkage, k=5

```
Number of clusters: 5
Transformation: Standardize (Z)
Method: Complete-linkage
Distance function: Euclidean

Cluster centers:
   | INCPERCAP|Und20  |Ov65   |Minrty |Unemprt|Noveh  |Lt_high|Rentocc|Rntburd|Noeng
-- |----------|-------|-------|-------|-------|-------|-------|-------|-------|-------
C1 |19837.7   |28.6244|13.6547|96.4962|18.6235|19.0193|16.4653|54.9558|35.6938|8.85518
C2 |35389.4   |22.8178|15.5076|56.4373|6.53725|8.46062|11.6261|36.2976|20.5243|36.3739
C3 |72820.9   |16.1342|9.23373|35.9495|3.32906|16.3783|4.45316|51.169 |19.6864|23.2746
C4 |19449.7   |17.2442|25.9271|93.9723|17.3231|23.7275|24.6656|51.3313|29.5007|27.0906
C5 |19511.1   |29.6646|10.5534|88.8032|10.5186|7.20325|30.7267|43.4111|24.971 |72.5091

The total sum of squares:  760
Within-cluster sum of squares:
   |Within cluster S.S.
-- |-------------------
C1 |85.0895
C2 |147.629
C3 |29.4238
C4 |14.3004
C5 |27.0794

The total within-cluster sum of squares:   303.522
The between-cluster sum of squares:    456.478
The ratio of between to total sum of squares:  0.600629
```

Figure 5.23: Summary – Complete Linkage, k=5

The summary in Figure 5.23 reveals similar performance but slightly inferior to Ward's method. The ratio of between to total is 0.600629, compared to 0.644089 for Ward's method.

The cluster map in Figure 5.24 shows how several of the previously identified outliers are now grouped in cluster 5. The overall spatial layout is similar to that obtained for Ward's method, although the cluster sizes are different.

5.4.5   Average linkage

Finally, the average linkage criterion suffers from some of the same problems as single linkage, although it yields slightly better results. The dendrogram in Figure 5.25 indicates a

Figure 5.24: Cluster Map – Complete Linkage, k=5

Figure 5.25: Dendrogram – Average Linkage, k=5

somewhat unbalanced structure, with two singletons and two clusters much larger than the third.

As given in Figure 5.26, the summary characteristics are better than in the single linkage case, with only two singletons. However, the overall ratio of between to total sum of squares is still considerably inferior to Ward's and complete linkage, at 0.522972.

The cluster map in Figure 5.27 again identifies two of the outliers from the single linkage map as singletons, but the others are absorbed into the larger clusters.

```
Number of clusters: 5
Transformation: Standardize (Z)
Method: Average-linkage
Distance function: Euclidean

Cluster centers:
   |INCPERCAP|Und20  |Ov65   |Minrty |Unemprt|Noveh  |Lt_high|Rentocc|Rntburd|Noeng
 --|---------|-------|-------|-------|-------|-------|-------|-------|-------|-------
 C1|11857    |44.2783|4.7975 |98.7904|30.4174|30.3275|15.6869|76.9207|30.7093|3.09423
 C2|21858    |18.5395|24.8198|86.0494|8.20697|19.9246|37.1823|61.1907|27.2874|69.4732
 C3|60754.3  |18.4095|11.6548|36.8351|3.73603|14.5189|5.42139|45.2165|19.3035|23.5778
 C4|22292    |25.8532|16.065 |95.225 |17.2127|17.5238|15.1728|49.5939|33.2613|8.23925
 C5|25780.7  |26.6328|12.7058|70.9604|8.08595|6.52803|22.3837|39.6367|22.5407|61.6839

The total sum of squares:   760
Within-cluster sum of squares:
   |Within cluster S.S.
 --|--------------------
 C1|0
 C2|0
 C3|102.457
 C4|137.427
 C5|122.658

The total within-cluster sum of squares:   362.542
The between-cluster sum of squares:   397.458
The ratio of between to total sum of squares:  0.522971
```

Figure 5.26: Summary – Average Linkage, k=5

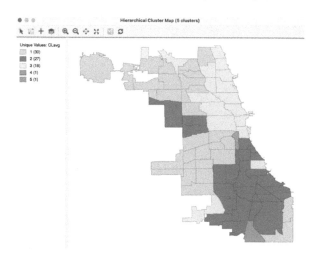

Figure 5.27: Cluster Map – Average Linkage, k=5

5.4.6   Sensitivity Analysis

Many parameters can be altered to assess the sensitivity of the cluster solution. This does not only pertain to the number of clusters (k) and the linkage method, but also to the distance metric, and, to a lesser extent, to the transformation.

The advantage of hierarchical clustering is that it is straightforward to change the number of clusters and assess the effect on cluster characteristics and spatial layout. As mentioned, a drawback is that observations are locked into a particular branch of the dendrogram, which limits the flexibility of later grouping.

The resulting clusters can be exploited in a number of different data analyses, which is pursued further in the next chapter.

6

Partitioning Clustering Methods

The principle behind *partitioning* clustering methods is to assign each observation to one out of k clusters. The objective is the same as that of hierarchical clustering, except that the number of clusters (traditionally referred to as k) is pre-specified. For a given k, the optimal assignment of observations to clusters is obtained such that, as before, the similarity within each cluster and the dissimilarity between clusters are maximized.

In this chapter, **K-means** clustering is covered, the most familiar example of a partitioning method. More advanced partitioning methods are considered in Chapter 7.

The earliest solution of the K-means problem is commonly attributed to Lloyd (1982) and referred to as *Lloyd's algorithm* (the algorithm was first contained in a Bell Labs technical report by Lloyd from 1957). It implements a so-called *iterative relocation*, i.e., the gradual improvement from an initial solution obtained by moving (relocating) observations to a different cluster. The K-means algorithm can be considered to be a special case of the EM (expectation-maximization) algorithm of Dempster et al. (1977). The expectation step then consists of allocating each observation to its nearest cluster center, and the maximization step is the recalculation of those cluster centers for each new layout (Han et al., 2012). While the progression of the iterative relocation is fairly straightforward, the choice of the initial starting point is not. It requires a careful sensitivity analysis. An extensive evaluation of K-means on a number of benchmark data sets is contained in Fränti and Sieranoja (2018).

The chapter begins with a step-by-step illustration of the K-means algorithm. A discussion of initial starting points leads to coverage of the K-means++ approach. The implementation in **GeoDa** is outlined, with a discussion of the interpretation, visualization and *spatialization* of cluster results, options and sensitivity analysis, and the use of cluster categories as variables.

The K-means method is illustrated with a subset of data on socio-economic determinants of health for 791 census tracts in Chicago contained in the *Chicago SDOH* sample data set.

6.1 Topics Covered

- Understand the principles behind K-means clustering
- Know the requirements to carry out K-means clustering
- Combine dimension reduction and cluster analysis
- Interpret the characteristics of a cluster analysis
- Assess the spatial representation of the clusters
- Carry out a sensitivity analysis to various parameters
- Impose a bound on the clustering solutions
- Use an elbow plot to pick the best k
- Use the cluster categories as a variable

DOI: 10.1201/9781032713175-6

GeoDa Functions

- Clusters > K Means
  - select variables
  - select K-means starting algorithms
  - select standardization methods
  - K-means characteristics
  - mapping the clusters
  - changing the cluster labels
  - saving the cluster classification
  - setting a minimum bound
- Explore > Conditional Plot > Box Plot
- Table > Aggregate
- Tools > Dissolve

Toolbar Icons

Figure 6.1: Clusters > K Means | K Medians | K Medoids | Spectral | Hierarchical

6.2　The K Means Algorithm

As in the case of hierarchical clustering, the point of departure is the sum of dissimilarities between all pairs of observations, which can be separated into a component for within dissimilarity and a component for between dissimilarity:

$$T = (1/2)(\sum_{h=1}^{k}[\sum_{i \in h}\sum_{j \in h}d_{ij} + \sum_{i \in h}\sum_{j \notin h}d_{ij}]) = W + B.$$

As before, W and B are complementary, so the lower W, the higher B, and vice versa.

Partitioning methods differ in terms of how the dissimilarity d_{ij} is defined and how the term W is minimized. Complete enumeration of all the possible allocations is unfeasible except for toy problems. There is no analytical solution.

The problem is NP-hard, so the solution has to be approached by means of a heuristic, as an iterative descent process. This is accomplished through an algorithm that changes the assignment of observations to clusters so as to improve the objective function at each step. All feasible approaches are based on what is called *iterative greedy descent*. A greedy algorithm is one that makes a locally optimal decision at each stage. It is therefore not guaranteed to end up in a *global* optimum, but may get stuck in a *local* one instead. The optimization strategy is based on an *iterative relocation* heuristic.

The K-means algorithm uses the squared Euclidean distance as the measure of dissimilarity:

$$d_{ij}^2 = \sum_{v=1}^{p} (x_{iv} - x_{jv})^2 = ||x_i - x_j||^2,$$

where the customary notation has been adjusted to designate the number of variables/dimensions as p instead of k (since k is traditionally used for the number of clusters).

This gives the overall objective as finding the allocation $C(i)$ of each observation i to a cluster h out of the k clusters so as to minimize the within-cluster similarity over all k clusters:

$$\min(W) = \min(1/2) \sum_{h=1}^{k} \sum_{i \in h} \sum_{j \in h} ||x_i - x_j||^2,$$

where, in general, x_i and x_j are p-dimensional vectors.

A little bit of algebra shows how this simplifies to minimizing the squared difference between the values of the observations in each cluster and the corresponding cluster mean:

$$\min(W) = \min \sum_{h=1}^{k} n_h \sum_{i \in h} (x_i - \bar{x}_h)^2.$$

In other words, minimizing the sum of (one half) of all squared distances is equivalent to minimizing the sum of squared deviations from the mean in each cluster, the *within* sum of squared errors.

Mathematical details

The objective function is one half the sum of the squared Euclidean distances for each cluster between all the pairs $i - j$ that form part of the cluster h, i.e., for all $i \in h$ and $j \in h$:

$$W = (1/2) \sum_{h=1}^{k} \sum_{i \in h} \sum_{j \in h} ||x_i - x_j||^2,$$

where, to keep the notation simple, the x is taken as univariate (without loss of generality).

For each i in a given cluster h, the sum of the squared distances to all the j in the cluster is:

$$\sum_{j} (x_i - x_j)^2 = \sum_{j} (x_i^2 - 2x_i x_j + x_j^2),$$

dropping the $\in h$ notation for simplicity.

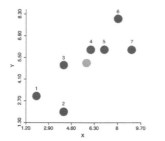

ID	X	Y	SSE
1	2	3	18.388
2	4	2	12.816
3	4	5	2.959
4	6	6	0.816
5	7	6	2.388
6	8	8	13.388
7	9	6	11.531
	5.714	5.143	62.286

Figure 6.2: K-means toy example   Figure 6.3: Worked example – basic data

With n_h observations in cluster h, $\sum_j x_i^2 = n_h x_i^2$ (since x_i does not contain the index j and thus is simply repeated n_h times). Also, $\sum_j x_j = n_h \bar{x}_h$, where $\bar{x}_h$ is the mean of x for cluster h. As a result:

$$\sum_j (x_i - x_j)^2 = n_h x_i^2 - 2n_h x_i \bar{x}_h + \sum_j x_j^2.$$

Next, using the same approach as for j, and since $\sum_j x_j^2 = \sum_i x_i^2$, the sum over all i becomes:

$$\sum_i \sum_j (x_i - x_j)^2 = n_h \sum_i x_i^2 - 2n_h^2 \bar{x}_h^2 + n_h \sum_i x_i^2 = 2n_h^2 (\sum_i x_i^2 / n_h - \bar{x}_h^2).$$

Finally, since $\sum_i x_i^2 / n_h - \bar{x}_h^2 = (1/n_h) \sum_i (x_i - \bar{x}_h)^2$ (the definion of variance), the contribution of cluster h to the objective function becomes:

$$(1/2) \sum_i \sum_j (x_i - x_j)^2 = n_h [\sum_i (x_i - \bar{x}_h)^2].$$

For all the clusters jointly, this is simply summed over h.

6.2.1 Iterative Relocation

The K-means algorithm is based on the principle of *iterative relocation*. In essence, this means that after an initial solution is established, subsequent moves (i.e., allocating observations to clusters) are made to improve the objective function. This is a greedy algorithm, that ensures that at each step the total within-cluster sums of squared errors (from the respective cluster means) is lowered. The algorithm stops when no improvement is possible. However, this does not ensure that a *global* optimum is achieved (for an early discussion, see Hartigan and Wong, 1979). Therefore sensitivity analysis is essential. This is addressed by trying many different initial allocations (typically assigned randomly).

To illustrate the logic behind the algorithm, the same simple toy example as in Chapter 5 is employed. The seven observations are shown in Figure 6.2, with the center (the mean of X and the mean of Y) shown in a lighter color. In K-means, the initial center is *not* one of the observations. The objective is to group the seven points into two clusters.

The coordinates of the seven points are given in the X and Y columns of Figure 6.3. The column labeled as SSE shows the squared distance from each point to the center of the point

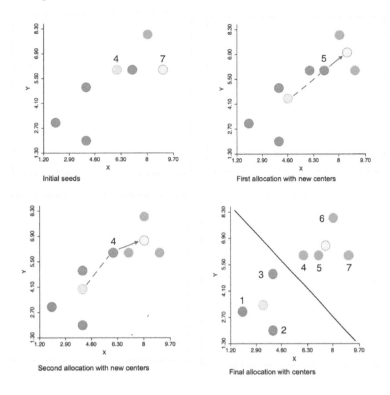

Figure 6.4: Steps in the K-means algorithm

ID	d_i4	d_i7
1	25	58
2	20	41
3	5	26
4	0	9
5	1	4
6	8	5
7	9	0

Figure 6.5: Squared distance to seeds

cloud, computed as the average of the X and Y coordinates (X=5.714, Y=5.143).[1] The sum of the squared distances constitutes the total sum of squared errors, or TSS, which equals 62.286 in this example. This value does not change with iterations, since it pertains to all the observations taken together and ignores the cluster allocations.

The initial step of the algorithm is to *randomly* pick two *seeds*, one for each cluster. The seeds are actual observations, not some other random location. In the example, observations 4 and 7 are selected, shown by the lighter color in the upper left panel of Figure 6.4. The other observations are allocated to the cluster whose seed they are closest to.

The allocation of observations to the cluster center they are closest to requires the distance from each point to the two *seeds*. The columns d_i4 and d_i7 in Figure 6.5 contain those respective distances. The highlighted values indicate how the first allocation consists of five

---

[1]Each squared distance is $(x_i - \bar{x})^2 + (y_i - \bar{y})^2$.

Cluster 1					Cluster 2			
ID	X	Y	SSE		ID	X	Y	SSE
1	20	41	8.72		6	8	8	1.25
2	5	26	6.12		7	9	6	1.25
3	0	9	0.72		Center	8.5	7	2.5
4	1	4	4.52					
5	8	5	8.32					
Center	6.8	17	28.4					

Figure 6.6: Step 1 – Summary characteristics

ID	d_i_C1	d_i_C2
1	8.72	58.25
2	6.12	45.25
3	0.72	24.25
4	4.52	7.25
5	8.32	3.25
6	24.52	1.25
7	21.92	1.25

Figure 6.7: Squared distance to Step 1 centers

Cluster 1					Cluster 2			
ID	X	Y	SSE		ID	X	Y	SSE
1	2	3	5		5	7	6	1.444
2	4	2	4		6	8	8	1.778
3	4	5	1		7	9	6	1.444
4	6	6	8		Center	8	6.667	4.667
Center	4	4	18					

Figure 6.8: Step 2 – Summary characteristics

ID	d_i_C1	d_i_C2
1	5	49.44
2	4	37.78
3	1	18.78
4	8	4.44
5	13	1.44
6	32	1.78
7	29	1.44

Figure 6.9: Squared distance to Step 2 centers

observations in cluster 1 (1-5) and two observations in cluster 2 (6 and 7). They are given blue and green colors in the upper left panel of Figure 6.4.

To each of the initial clusters corresponds a new central point, computed as the average of the respective X and Y coordinates, respectively (6.8, 17) and (8.5, 7). The SSE follows as the squared distance between each observation in the cluster and its central point, as listed in Figure 6.6. The sum of the total SSE in each cluster is the *within* sum of squares, WSS = 28.4 + 2.5 = 30.9 in this first step. Consequently, the *between* sum of squares BSS = TSS - WSS = 62.3 - 30.9 = 31.4. The associated ratio BSS/TSS, an indicator of the quality of the cluster, is 0.50.

With the new cluster center points in place, each observation is again allocated to the closest center. From the calculations in Figure 6.7, this results in observation 5 moving from cluster 1 to cluster 2, which now consists of three observations (cluster 1 has the remaining four). This move is illustrated in the upper right panel of Figure 6.4.

The new clusters yield cluster centers at (4, 4) and (8, 6.667). The associated SSE are shown in Figure 6.8. This results in an updated WSS of 18 + 4.667 = 22.667, clearly an improvement

Cluster 1				Cluster 2			
ID	X	Y	SSE	ID	X	Y	SSE
1	2	3	1.889	4	6	6	2.5
2	4	2	2.222	5	7	6	0.5
3	4	5	3.222	6	8	8	2.5
Center	3.333	3.333	7.333	7	9	6	2.5
				Center	7.5	6.5	8

Figure 6.10: Step 3 – Summary characteristics

of the objective function (down from 30.9). The corresponding ratio of BSS/TSS becomes 0.64, also a substantial improvement relative to 0.50.

Figure 6.9 lists the distances for each point to the new cluster centers. This results in observation 4 moving from cluster 1 to cluster 2, shown in the lower left panel of Figure 6.4.

The new cluster centers are (3.333, 3.333) and (7.5, 6.5). The associated SSE are shown in Figure 6.10, with a WSS of $7.333 + 8 = 15.333$ and an associated BSS/TSS of 0.75. This latest allocation no longer results in a change, yielding a local optimum, illustrated in the lower-right panel in Figure 6.4.

This solution corresponds to the allocation that would result from a Voronoi diagram or Thiessen polygon around the cluster centers. All observations are closer to their center than to any other center. This is indicated by the black line in the graph, which is perpendicular at the midpoint to an imaginary line connecting the two centers and separates the area into two compact *regions*.

6.2.2 The Choice of K

A key element in the K-means method is the choice of the number of clusters, k. Typically, several values for k are considered, and the resulting clusters are then compared in terms of the objective function.

Since the total sum of squared errors (SSE) equals the sum of the within-group SSE and the total between-group SSE, a common criterion is to assess the ratio of the total between-group sum of squares (BSS) to the total sum of squares (TSS), i.e., BSS/TSS. A higher value for this ratio suggests a better separation of the clusters. However, since this ratio increases with k, the selection of a *best* k is not straightforward. Several ad hoc rules have been suggested, but none is totally satisfactory.

One useful approach is to plot the objective function against increasing values of k. This could be either the within sum of squares (WSS), a decreasing function with k or the ratio BSS/TSS, a value that increases with k. The goal of a so-called *elbow plot* is to find a *kink* in the progression of the objective function against the value of k (see Section 6.3.4.2). The rationale behind this is that as long as the optimal number of clusters has not been reached, the improvement in the objective should be substantial, but as soon as the optimal k has been exceeded, the curve flattens out. This is somewhat subjective and often not that easy to interpret in practice. GeoDa's functionality does not include an elbow plot, but all the information regarding the objective functions needed to create such a plot is provided in the output. This is illustrated in Section 6.3.4.2.

A more formal approach is the so-called Gap statistic of Tibshirani et al. (2001), which employs the difference between the log of the WSS and the log of the WSS of a uniformly randomly generated reference distribution (uniform over a hypercube that contains the data) to construct a *test* for the optimal k. Since this approach is computationally quite demanding, it is not further considered.

6.2.3   K-means++

Typically, the assignment of the first set of k cluster centers is obtained by uniform random sampling k distinct observations from the full set of n observations. In other words, each observation has the same probability of being selected.

The standard approach is to try several random assignments and start with the one that gives the best value for the objective function (e.g., the smallest WSS, or the largest BSS/TSS). This is one of the two approaches implemented in `GeoDa`. In order to ensure that the results are reproducible, it is important to set a seed value for the random number generator. Also, to further assess the sensitivity of the result to the starting point, different seeds should be tried (as well as a different number of initial solutions).[2]

A second approach uses a careful consideration of initial seeds, following the procedure outlined in Arthur and Vassilvitskii (2007), commonly referred to as **K-means++**. The rationale behind K-means++ is that rather than sampling uniformly from the n observations, the probability of selecting a new cluster seed is changed in function of the distance to the nearest existing seed. Starting with a uniformly random selection of the first seed, say c_1, the probabilities for the remaining observations are computed as:

$$p_{j \neq c_1} = \frac{d_{jc_1}^2}{\sum_{j \neq c_1} d_{jc_1}^2}.$$

In other words, the probability is no longer uniform, but changes with the squared distance to the existing seed: the smaller the distance, the smaller the probability. This is referred to by Arthur and Vassilvitskii (2007) as the squared distance (d^2) weighting.

The weighting increases the chance that the next seed is further away from the existing seeds, providing a better coverage over the support of the sample points. Once the second seed is selected, the probabilities are updated in function of the new distances to the closest seed and the process continues until all k seeds are picked.

While generally being faster and resulting in a superior solution in small to medium-sized data sets, this method does not scale well (as it requires k passes through the whole data set to recompute the distances and the updated probabilities). Also, a choice of a large number of random initial allocations may yield a better outcome than the application of K-means++, at the expense of a somewhat longer execution time.

6.3   Implementation

The K-means algorithm is illustrated with a partial replication of the cluster analysis in Kolak et al. (2020). The variables selected are matched as closely as possible, but instead of using all the U.S. census tracts, the analysis is based on the 791 Chicago census tracts in the *Chicago SDOH* sample data set. The 16 variables are listed in Figure 6.11 with some descriptive statistics. Their abbreviations are given below:

---

[2] `GeoDa` implements this approach by leveraging the functionality contained in the *C Clustering Library* (de Hoon et al., 2017).

Variable	Abbrev	Mean	St.Dev	Min	Max
Minority population share	EP_MINRTY	69.5	30.3	6.2	100.0
Population share aged > 65	Ovr6514P	10.7	6.0	0.7	47.0
Population share aged < 18	EP_AGE17	22.6	8.5	1.1	50.7
Disabled population share	EP_DISABL	11.3	5.5	0.4	34.4
No high school	EP_NOHSDP	19.2	13.6	0.0	65.4
Limited English profiency	EP_LIMENG	7.8	9.9	0.0	47.0
Percent single parent households	EP_SNGPNT	13.0	9.8	0.0	56.5
Poverty rate	Pov14	20.1	14.8	0.0	73.1
Per capita income	PerCap14	27957	19592	3077	127743
Unemployment rate	Unemp14	15.6	10.1	0.6	51.8
Percent without health insurance	EP_UNINSUR	18.5	9.1	1.0	45.0
Percent crowded housing	EP_CROWD	4.8	4.8	0.0	25.8
No car	EP_NOVEH	26.9	14.9	0.0	79.2
Child poverty rate	ChildPvt14	31.1	22.4	0.0	93.6
Health literacy index	HealthLit	243	14.6	126	272
Foreclosure risk	FORCLRISK	23.5	23.8	0.2	100.0

Figure 6.11: SDOH Census Tract Variables Descriptive Statistics

- EP_MINRTY: minority population share
- Ovr6514P: population share aged > 65
- EP_AGE17: population share aged < 18
- EP_DISABL: disabled population share
- EP_NOHSDP: no high school
- EP_LIMENG: limited English proficiency
- EP_SNGPNT: percent single parent households
- Pov14: poverty rate
- PerCap14: per capita income
- Unemp14: unemployment rate
- EP_UNINSUR: percent without health insurance
- EP_CROWD: percent crowded housing
- EP_NOVEH: no car
- ChildPvt14: child poverty rate
- HealtLit: health literacy index
- FORCLRISK: foreclosure risk

This set nearly matches the variables listed in Table 2 of Kolak et al. (2020), except that for the Chicago census tract data, no variables were included for Renter and Rent Burden. Instead, the variables for child poverty rate, health literacy index and foreclosure risk were added. Most variables are characterized by a large range and associated standard deviation. A sense of the effect of the spatial scale, i.e., census tract compared to community area, can be gleaned from a comparison of the descriptive statistics for matching variables in Figure 5.13.

In practice, when many variables are considered, it is often more effective to carry out the clustering exercise after dimension reduction. Instead of using the 16 variables listed, the clustering is based on the main principal components.[3]

---

[3]This follows the logic in the Kolak et al. (2020) paper. The difference is that the principal components are derived from the Chicago census tract observations, whereas in Kolak et al. (2020) they were computed for all U.S. census tracts.

```
Variable Loadings:
                   PC1          PC2          PC3          PC4
EP_MINRTY       0.338249   -0.0692012     0.129968    -0.0648413
Ovr6514P       -0.0276788   -0.321223      0.601939     0.198585
EP_AGE17        0.291592     0.107457    -0.0782367    -0.271201
EP_DISABL       0.195323    -0.363332      0.307376     0.197528
EP_NOHSDP       0.284731     0.286737      0.173658     0.12097
EP_LIMENG       0.101775     0.497496      0.233597     0.178097
EP_SNGPNT       0.316812    -0.0762848    -0.219022    -0.16748
Pov14           0.326665    -0.106628     -0.268768     0.111728
PerCap14       -0.322455    -0.0434401    -0.185723     0.0618045
Unemp14         0.287193    -0.258054    -0.0522028    -0.0839096
EP_UNINSUR      0.25713      0.29145       0.106432     0.167606
EP_CROWD        0.223664     0.33242     -0.0310081     0.174806
EP_NOVEH        0.115801    -0.295039     -0.368025     0.492778
ChldPvt14       0.317626    -0.0989162    -0.202578     0.0811426
HealthLit      -0.017049    -0.020497      0.0698693    0.547322
FORCLRISK       0.21436     -0.200962      0.29048     -0.372054
```

Figure 6.12: Principal Components Variable Loadings

6.3.1   Digression: Clustering with Dimension Reduction

A principal component analysis of the 16 variables yields four main components, according to the Kaiser criterion (see Section 2.3.2.1). As it turns out, this matches the results in Kolak et al. (2020) for the whole U.S.

The first component explains 44.3 % of the variance, the second 17.7 %, the third 8.6 %, and the fourth 6.7 %, for a total of 77.4 %. The principal loadings of the variables for the four components are listed in Figure 6.12.

The interpretation of the components is not that straightforward. Following Kolak et al. (2020), the main focus is on loadings that are larger than 0.30 in absolute value. Similar to the national results, the first component can be labeled *economic disadvantage*, with strong positive loadings on percent minority, poverty rate (including child poverty) and single-parent households, and a negative loading on income per capita.

The results for the second component also closely match the national findings, with strong positive loadings on limited English proficiency, crowded housing, and lack of health insurance, with negative loadings on disability, elderly and lack of vehicles. This would suggest more *immigrant* neighborhoods.

The loadings for the third and fourth components differ somewhat from the national results, which is not surprising, given the urban nature of the data, in contrast to a mix of urban and rural in the national sample.

For the third component, the strongest positive loadings are age over 65 and percent disabled, but with a negative loading on lack of cars, suggesting neighborhoods with an *older population*, further from the urban core, requiring car use (early suburbs). Finally, the fourth component loads strongly positive on health literacy and the lack of vehicles, and negative on foreclosure risk, suggesting better-educated *stable* neighborhoods with good transit access.

6.3.2   Cluster Parameters

K-Means clustering is invoked from the drop-down list associated with the cluster toolbar icon, as the first item in the classic clustering methods subset, shown in Figure 6.1. It can also be selected from the menu as **Clusters > K Means**.

This brings up the **KMeans Clustering Settings** dialog. This dialog has a similar structure to the one for hierarchical clustering, with a left-hand panel to select the variables and specify the cluster parameters, and a right-hand panel that lists the **Summary** results.

Figure 6.13: K-Means Clustering Variable Settings

The left panel is shown in Figure 6.13.

In the example, the four principal components are selected as variables, and the **Number of Clusters** is set to **8**. In contrast to hierarchical clustering, this is not set to a default value and must be specified. All other items are left to their default settings. This means that the **Initialization Method** is **KMeans++**, with **150 Initialization Re-runs**, and **1000 Maximum Iterations**. The random seed is set to the default value, but that can be altered. The variables are transformed to **Standardize (Z)**. Since K-means clustering only works for Euclidean distance, the **Distance Function** is not available.

Clicking on **Run** saves the cluster classification in the table under the field specified as such (here, **CLkm8**). It also provides the cluster characteristics in the **Summary** panel, with the associated cluster map in a separate window.

6.3.3 Cluster Results

The cluster characteristics reported in the **Summary** panel are the same as for hierarchical clustering. In fact, all cluster methods in GeoDa share the same summary format.

Figure 6.14 lists the results for K-means with $k=8$ on the four principal components of socio-economic determinants of health. The cluster centers are again reported, but these are not that easy to interpret, since they pertain to principal components, not to the original variables.

The total sum of squares is 3160. The within sum of squares yields 995.227, resulting in a between to total ratio of 0.685.

```
Method: KMeans
Number of clusters: 8
Initialization method: KMeans++
Initialization re-runs: 150
Maximum iterations: 1000
Transformation: Standardize (Z)
Distance function: Euclidean
```

Cluster centers:

	PC1	PC2	PC3	PC4
C1	2.98676	-1.25559	-1.08884	-0.0520222
C2	2.09779	2.73457	0.390127	0.310579
C3	-1.13147	0.672746	0.491659	0.2925
C4	-3.53033	0.195898	-0.978708	-0.393634
C5	-1.09521	0.450725	0.529864	-1.36672
C6	1.23994	-1.97944	1.41385	-0.631278
C7	-2.54594	-0.453212	-0.980015	1.30483
C8	-1.10457	-1.94505	1.34694	1.81058

```
The total sum of squares:   3160
Within-cluster sum of squares:
```

	Within cluster S.S.
C1	241.165
C2	162.102
C3	104.201
C4	89.1439
C5	124.588
C6	103.245
C7	69.8302
C8	100.952

```
The total within-cluster sum of squares:   995.227
The between-cluster sum of squares:   2164.77
The ratio of between to total sum of squares:  0.685055
```

Figure 6.14: Summary – K-Means Method, k=8

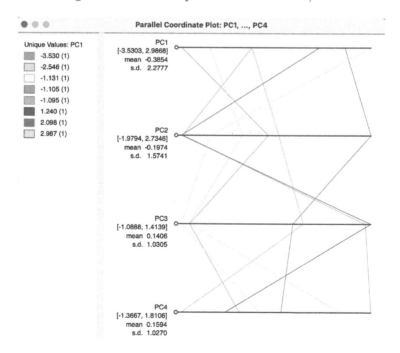

Figure 6.15: PCP Plot of K-Means Cluster Centers

Figure 6.15 visualizes the relative scores for the mean centers on each of the principal components for the eight clusters in a parallel coordinate plot. The colors associated with each cluster match the legend in the cluster map in Figure 6.16. Given the way the PCP

Figure 6.16: Cluster Map – K-Means Method, k=8

is implemented in `GeoDa`, the unique values classification is based on the first principal component. Even though the components are continuous variables, the cluster center scores are distinct, so that they can be treated as unique values.[4] Cluster 1 starts with the highest score for PC1, but then achieves low scores for the other components. Cluster 2 starts with the second-highest score for PC1 and the highest score for PC2, with middle-of-the-road values for the other components. Each of the polylines yields a clearly distinct pattern from the others, corresponding to the *dissimilarity* between clusters. With some imagination and creativity, the clusters can be associated with substantive concepts, such as economic distress for the first cluster. As with the interpretation of principal components, this is not always straightforward.

The spatial layout of the clusters is shown in Figure 6.16. The clusters range in size from 154 tracts for the largest to 44 for the smallest. They tend to consist of large chunks of contiguous tracts, which is likely a result of the high degree of racial and socio-economic spatial stratification in Chicago. A priori, one would not expect such spatially grouped results. For example, the largest cluster is made up predominantly by a set of contiguous tracts in the west and similarly in the south, corresponding with traditionally economically distressed (predominantly minority) neighborhoods. Cluster 4 (114 tracts) and cluster 7 (60 tracts) are primarily in northern lakeside of the city, well-known to consist of higher-income (predominantly white) neighborhoods.

In sum, even though no spatial constraints were imposed on the clustering method, the results bring up very pronounced (and well-known) spatial patterns in the cluster formations.

6.3.3.1 Adjusting cluster labels

Since the cluster map is simply a special case of a `GeoDa` unique values map, its legend categories can be moved around. This can be particularly useful when comparing the maps for different cluster classifications, where the ordering by size of cluster may not be the same.

---

[4]The graph is constructed by saving the cluster summary to a text file and subsequently extracting the mean center information. This then becomes the input to `GeoDa` as a csv file with the principal components as variables and the cluster labels as observations. The colors in the legend are edited to match the cluster map classification.

k	WSS	Ratio	Change	Percent
1	3160.00	0.000		
2	2506.41	0.207	0.207	
3	1955.03	0.381	0.174	84.1
4	1576.68	0.501	0.120	31.5
5	1317.64	0.583	0.082	16.4
6	1169.41	0.630	0.047	8.1
7	1073.75	0.660	0.030	4.8
8	995.23	0.685	0.025	3.8
9	916.92	0.710	0.025	3.6
10	856.08	0.729	0.019	2.7
11	804.35	0.745	0.016	2.2
12	759.16	0.760	0.015	2.0
13	728.64	0.769	0.009	1.2
14	698.79	0.779	0.010	1.3
15	674.75	0.786	0.007	0.9
16	647.31	0.795	0.009	1.1
17	625.73	0.802	0.007	0.9
18	604.04	0.809	0.007	0.9
19	583.88	0.815	0.006	0.7
20	564.21	0.821	0.006	0.7

Figure 6.17: K-Means – Evolution of fit with k

Changing the color (and thus the label) of the clusters by moving the legend rectangles up or down in the list may facilitate visual comparison of the cluster maps.

6.3.4 Options and Sensitivity Analysis

The K-means run with the default options yields a local optimum among several possible solutions. In order to assess the quality of the solution, several parameters and options can be manipulated. Most important of these are the initialization (the starting point for the iterative relocation) and the number of clusters, k. In addition, while Z standardization is often the norm, this is by no means a requirement. Some other approaches, which may be more robust to the effect of outliers (on the estimates of mean and variance used in the standardization) are the MAD, mentioned in the context of PCA, as well as range standardization. The latter is often recommended for cluster exercises (e.g., Everitt et al., 2011). It rescales each variable such that its minimum is zero and its maximum becomes one by subtracting the minimum and dividing by the range. For each cluster method, the usual full range of standardization options is available.

Finally, in many contexts, small clusters are undesirable and some minimum size constraint may need to be imposed. For example, this is often the case when reporting rates for rare diseases, where the population at risk must meet a minimum size (e.g., 100,000). GeoDa provides an option to set a minimum bound for the cluster size.

6.3.4.1 Initialization

The **Initialization Method** option provides **Random** initialization as an alternative to KMeans++. The number of initial solutions that are evaluated is set in the **Initialization Re-runs** box. Typically, the default of 150 is fine in most applications. In fact, this example, changing this option only affects the ultimate solution in marginal ways. The number of tracts in some clusters differs slightly, but several different combinations all yield a BSS/TSS of 0.685.

Changing this option is more important when the initial default results do not indicate a good separation between clusters.

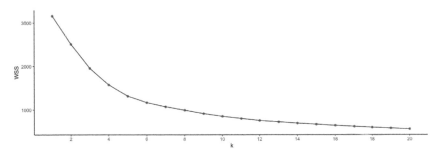

Figure 6.18: K-Means – Elbow Plot WSS

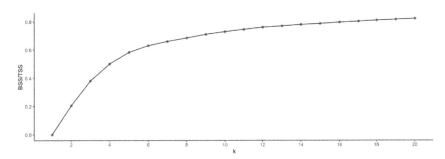

Figure 6.19: K-Means – Elbow Plot BSS/TSS

6.3.4.2   Selecting k – Elbow plot

Selecting the *optimal* k is critical both for the quality of the ultimate solution and for its interpretation. Unfortunately, this is not that easy in practice. An *elbow* plot shows the change in performance as k increases. By construction, the WSS will decrease and the BSS/TSS ratio will increase. The goal is to identify a *kink* in the plot, i.e., a value for k beyond which there are no more major improvements in the cluster characteristics.

As such, GeoDa does not include functionality for an elbow plot, but all the needed information is contained in the summary output. Figure 6.17 lists the values for the within sum of squares (WSS) and the ratio of between sum of squares to total sum of squares (BSS/TSS) for k going from 1 (when WSS=TSS) to 20. The third and fourth columns list the change in BSS/TSS as well as the percentage change.

The improvement in fit is quite substantial in the early stages, but it seems to stabilize between k=8 and k=10, moving to a much smaller change. For example, the change in value for BSS/TSS achieved between k=5 and k=8 (yielding 0.685) is about the same as that between k=8 and k=15 (yielding 0.786). Therefore, a choice of k between 8 and 10 seems to be reasonable in this application.

A graphical representation of the evolution of the fit is given in the elbow plots in Figures 6.18 and 6.19, respectively for WSS and for BSS/TSS. Whereas it is clear that the improvement in fit gradually decreases with higher k, identifying a clear *kink* in the curve is not that straightforward.

6.3.4.3   Standardization

In this example, the principal components are already somewhat standardized. By construction, the means are zero and the range is limited by the standardization on the factor

```
Method: KMeans
Number of clusters: 8
Initialization method: KMeans++
Initialization re-runs: 150
Maximum iterations: 1000
Minimum bound: 200000(Pop2014)
Transformation: Standardize (Z)
Distance function: Euclidean

Cluster centers:
   |PC1       |PC2       |PC3       |PC4
-- |----------|----------|----------|-----------
C1 |-3.36093  |0.0900402 |-1.066    |-0.00694963
C2 |-0.340233 |1.19822   |0.480225  |0.185436
C3 |2.69917   |-1.45174  |-0.643319 |0.590515
C4 |-1.74477  |0.275209  |0.515962  |-1.27858
C5 |2.41144   |3.01362   |0.388348  |0.398214
C6 |1.01095   |-1.94674  |1.5192    |-0.672525
C7 |3.0417    |-0.992954 |-1.35914  |-0.728996
C8 |-1.75941  |-1.31045  |0.758438  |1.70776

The total sum of squares:  3160
Within-cluster sum of squares:
   |Within cluster S.S.
-- |-------------------
C1 |151.865
C2 |118.198
C3 |123.098
C4 |135.516
C5 |115.716
C6 |93.8841
C7 |106.096
C8 |161.12

The total within-cluster sum of squares:  1005.49
The between-cluster sum of squares:    2154.51
The ratio of between to total sum of squares:  0.681806
```

Figure 6.20: Summary – K-Means Method, population bound 200,000, k=8

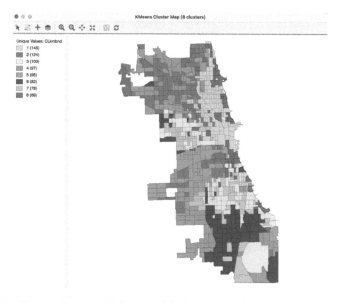

Figure 6.21: Cluster Map – K-Means Method, population bound 200,000, k=8

loadings. While the variance for each component is not identical, the difference in range is limited. Therefore, in this instance, different standardizations do not yield meaningfully different results. However, in general, depending on the context, even though the standard Z transformation is appropriate in most circumstances, there may be instances where a different transformation (or even no transformation) may be a better option.

6.3.4.4 Minimum bound

In several applications, it is necessary that each cluster achieves a minimum size, such as the population at risk in epidemiological studies or a critical population threshold in marketing studies. In GeoDa, it is possible to *constrain* the K-means clustering solution by imposing a

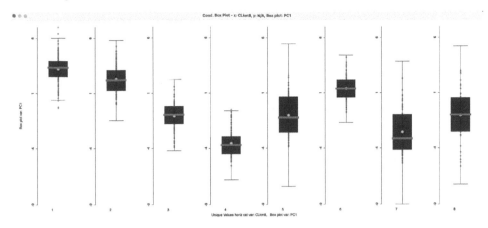

Figure 6.22: Conditional Box Plot by Cluster Category

minimum value for a spatially extensive variable, such as a population total. This ensures that the clusters meet a minimum size for that variable.

This is accomplished through the **Minimum Bound** option. It is turned on by checking the corresponding box in the left-hand panel of the interface and specifying a *spatially extensive* variable from the drop-down list. More specifically, the variable should be a count that lends itself to aggregation by means of a simple sum, rather than percentages, per capita ratios or median values.

For example, the variable **Pop2014** contains the population size for each tract. The default setting is to take the value of the 10th percentile for this variable, in this case a value of 271,050. However, with this value and for $k=8$, none of the solutions achieves the minimum bound (see also Section 6.4.2).

Instead, with a population threshold of 200,000, the results shown in Figures 6.20 and 6.21 are obtained. The *price* for imposing the constraint as opposed to an unconstrained solution is a slight deterioration in the BSS/TSS ratio, to 0.681806.

Relative to the unconstrained solution, the cluster sizes are slightly more balanced, with somewhat smaller larger clusters and somewhat larger smaller clusters. Also, as the cluster map illustrates, the location of the larger clusters has shifted, from predominantly in the south and west for the unconstrained solution to a northern location for the constrained solution. Other such spatial shifts can be observed as well.

6.4 Cluster Categories as Variables

The cluster categories are saved to the Table as an integer variable. This variable can be used to define categories in a bubble chart or in conditional plots, or, more formally, in an analysis of variance. In addition, it can form the basis of an aggregation or dissolve operation on the original data.

Figure 6.23: Dissolved Cluster Categories

	CLkm8	AGG_COUNT	Pop2014
1	1	154	413926
2	2	125	519489
3	3	116	470395
4	4	114	354970
5	5	89	329598
6	6	89	285996
7	7	60	213285
8	8	44	122842

Figure 6.24: Aggregated Cluster Categories

6.4.1 Conditional Box Plot

An intuitive application of the cluster categories is to employ them as the conditioning variable in a conditional graph (see Volume 1). For example, the distribution for the first principal component can be compared among cluster categories in a conditional box plot, as shown in Figure 6.22. In contrast to Figure 6.15, where only the center mean for PC1 is shown for each cluster, the box plot shows the full range of the distribution as well as the mean and median in each cluster. The relative positions of the mean (green dot) match the pattern on the parallel coordinate plot for the mean centers. Taking the interquartile range into account suggests worse economic conditions in clusters 1, 2 and 6, whereas clusters 4 and 7 are clearly at the opposite end of the spectrum.

6.4.2 Aggregation by Cluster

The cluster category can also be used to create a new set of observations at the cluster scale. This can be done through simple aggregation or by means of the GIS dissolve operation (covered in detail in Chapter 3 of Volume 1). Whereas the former only results in a new data table, the latter yields a new spatial layer.

It is important to keep in mind what operation is used to compute the aggregated values. For spatially extensive variables, a simple sum is appropriate, but for other variables this is not the case.

The dissolve operation is illustrated in Figure 6.23. This shows essentially the same pattern as the cluster map, but the tract boundaries have been removed and only the spatial definition of the clusters remains.

The associated data table, for a **Sum** operation on the **Pop2014** variable is shown in Figure 6.24. The table includes the cluster label, the number of census tracts contained in each cluster (**AGG_COUNT**) and the total population of each cluster. The table reveals that clusters 7 (population 213,285) and 8 (population 122,842) do not meet the default minimum bounds population criterion of 271,050, thus confirming that the latter is too high to obtain a constrained solution.

The same table is obtained when the aggregation operation is carried out without the spatial dissolve.

7

Advanced Clustering Methods

This third chapter in the treatment of classic clustering methods considers two more advanced partitioning clustering techniques. The first, **K-Medians**, is a straightforward variant on K-Means. It operates in exactly the same manner, except that the representative *center* of each cluster is the *median* instead of the mean.

The second method is similar to K-Medians and is often confused with it. **K-Medoids** is a method where the center of each cluster is an actual cluster member, in contrast to what holds for both K-Means and K-Medians.

Both methods operate in largely the same manner as K-Means, but they differ in the way the central point of each cluster is defined and the manner in which the *nearest* points are assigned.

Each method will be covered in turn. However since their implementation shares a lot of commonalities with the methods covered in Chapters 5 and 6, the focus will be only on those aspects that differ.

The *Chicago SDOH* sample data set is again used to illustrate the techniques.

7.1 Topics Covered

- Understand the difference between K-Medians and K-Medoids clustering
- Carry out and interpret the results of K-Medians clustering
- Gain insight into the logic behind the PAM, CLARA and CLARANS algorithms for K-Medoids
- Carry out and interpret the results of K-Medoids clustering

GeoDa Functions

- Clusters > K Medians
- Clusters > K Medoids

7.2 K-Medians

K-Medians is a variant of K-Means clustering. As a partitioning method, it starts by randomly picking k starting points and assigning observations to the nearest initial point. After the assignment, the center for each cluster is re-calculated and the assignment process repeats

DOI: 10.1201/9781032713175-7

```
Method: KMedians
Number of clusters: 8
Initialization method: Random
Initialization re-runs: 150
Maximum iterations: 1000
Transformation: Standardize (Z)
Distance function: Manhattan

Cluster centers: (median)
   |PC1        |PC2       |PC3       |PC4
   |-----------|----------|----------|-----------
-- |-----------|----------|----------|-----------
C1 |3.10941    |-1.31582  |-0.935003 |-0.0648063
C2 |-3.81514   |0.149594  |-1.18328  |-0.0613904
C3 |-0.558074  |0.952879  |0.147842  |0.451371
C4 |-2.57926   |0.0147099 |0.461734  |-0.719024
C5 |1.15439    |-1.95528  |1.44314   |-0.691304
C6 |0.949734   |1.8457    |0.770973  |-0.623733
C7 |2.64427    |3.11014   |0.312014  |0.608475
C8 |-2.0366    |-1.39886  |0.617999  |1.753

The total sum of squares:   3160
Within-cluster sum of squares:
   |Within cluster S.S.
-- |--------------------
C1 |259.896
C2 |133.747
C3 |141.389
C4 |89.3498
C5 |120.025
C6 |73.5407
C7 |83.4122
C8 |160.554

The total within-cluster sum of squares:   1061.91
The between-cluster sum of squares:        2098.09
The ratio of between to total sum of squares:  0.663951

(Using Manhattan distance to medians)
The total sum of distance: 2555.97
Within-cluster sum of distances:
   |Within Cluster D |Average
-- |-----------------|-------
C1 |303.023          |1.89389
C2 |190.162          |1.53356
C3 |168.25           |1.57243
C4 |134.583          |1.47893
C5 |155.573          |1.7096
C6 |112.142          |1.45639
C7 |126.496          |1.66443
C8 |157.217          |2.41872

The total within-cluster sum of distance:  1347.45
The ratio of total within to total sum of distance: 0.527177
```

Figure 7.1: Summary – K-Medians Method, k=8

itself. In this way, K-Medians proceeds in exactly the same manner as K-Means (for specifics, see Section 6.2.1). It is in fact also an EM algorithm.

In contrast to K-Means, the central point is not the average (in multi-attribute space), but instead the *median* of the cluster observations. The median center is computed separately for each dimension, so it is typically not an actual observation (similar to what is the case for the cluster average in K-Means).

The objective function for K-Medians is to find the allocation $C(i)$ of observations i to clusters $h = 1, \ldots k$, such that the sum of the *Manhattan distances* between the members of each cluster and the cluster median is minimized:

$$\mathrm{argmin}_{C(i)} \sum_{h=1}^{k} \sum_{i \in h} ||x_i - x_{h_{med}}||_{L_1},$$

where the distance metric follows the L_1 norm, i.e., the Manhattan block distance.

K-Medians is often confused with K-Medoids. However, there is an important difference in that in K-Medoids, the central point has to be one of the observations (Kaufman and Rousseeuw, 2005). K-Medoids is considered in Section 7.3.

The Manhattan distance metric is used to assign observations to the nearest center. From a theoretical perspective, this is superior to using Euclidean distance since it is consistent with the notion of a median as the center (de Hoon et al., 2017, p. 16).

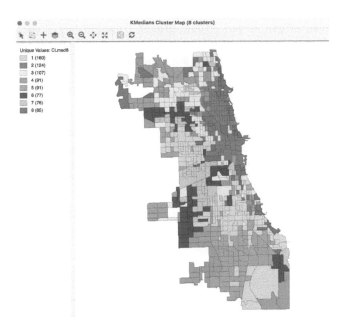

Figure 7.2: Cluster Map – K-Medians Method, k=8

In most other respects, the implementation and interpretation of K-Medians is the same as for K-Means. There is one important difference regarding the measure of fit of the cluster assignment. Since the reference point is no longer the mean, but instead refers to the median, the sum of squared deviations from the mean is not a very meaningful metric. Instead, one can assess the within cluster *Manhattan distance* from all cluster members to the cluster median. The total of such distances across all clusters can be related to the total distance to the median of all the data points. A *smaller* ratio of within to total distance is an indication of a better cluster assignment.

7.2.1 Implementation

K-Medians clustering is invoked from the drop-down list associated with the cluster toolbar icon, as the second item in the classic clustering methods subset, part of the full list shown in Figure 6.1. It can also be selected from the menu as **Clusters > K Medians**.

This brings up the **KMedians Clustering Settings** dialog. This dialog has an almost identical structure to the one for K-Means clustering (see Figure 6.13), with a left-hand panel to select the variables and specify the cluster parameters, and a right-hand panel that lists the **Summary** results.

The same example is used as in Chapter 6, with the variables as the first four principal components computed from 16 socio-economic determinants of health for 791 Chicago census tracts (see Section 6.3.1).

With all the parameters set to their default values and $k = 8$, the summary characteristics are listed in Figure 7.1. The results are organized in the same manner as for other clustering methods, starting with a brief list of the method and parameter values, followed by the cluster centers. In contrast to K-Means, the cluster centers for K-Medians are computed as the median of the observations in multi-attribute space that belong to each cluster. They are interpreted in the same way as before.

The next table of results is not that meaningful. It consists of the usual sum of squares metrics, but these are not really an indicator of how well the algorithm performed, since they are based on a different objective function. They remain however useful to compare the results to those of other methods. The BSS/TSS ratio is 0.663951, somewhat inferior to the K-Means result.

The next table of distance measures is the one actually used in the objective function. For each cluster, the total is listed of the Manhattan distances of the cluster members to the cluster median, as well as their average (all else being the same, clusters with more members would have a higher within cluster distance). This reveals that the *tightest* cluster is C6, with an average distance of 1.45639. In contrast, clusters C8 (2.41872) and C1 (1.89389) are much less compact. The total sum of the Manhattan distances to the median of the data set is 2555.97. The total within cluster distances amounts to 1347.45, yielding a ratio of within to total of 0.527177 (the lower the ratio, the better).

The spatial layout of the results is shown in the cluster map in Figure 7.2. The overall pattern shows great similarity with the K-Means cluster map in Figure 6.16, although the cluster sizes differ slightly. The largest cluster now consists of 160 tracts (compared to 154), and the smallest of 65 (compared to 44). While detailed cluster membership differs in minor ways, the overall pattern of where the clusters are located in the city is very similar.

7.2.2 Options and Sensitivity Analysis

K-Medians has the same options as K-Means with respect to variable transformations, maximum iterations and minimum bounds (see Section 6.3.4). KMeans++ initialization is not available, nor is the **Distance Function**, since only Manhattan distance is used.

As before, the cluster categories are saved as an integer variable, which can be used in other analyses.

7.3 K-Medoids

The objective of the K-Medoids algorithm is to minimize the sum of the distances from the observations in each cluster to a *representative center* for that cluster. In contrast to K-Means and K-Medians, the cluster centers are actual observations and thus do not need to be computed separately (as mean or median).

K-Medoids works with any dissimilarity matrix, since it only relies on inter-observation distances and does not require a mean or median center. When actual observations are available, the Manhattan distance is the preferred metric, since it is less affected by outliers. In addition, since the objective function is based on the sum of the actual distances instead of their squares, the influence of outliers is even smaller.[1]

The objective function is expressed as finding the cluster assignments $C(i)$ such that, for a given k:

$$\text{argmin}_{C(i)} \sum_{h=1}^{k} \sum_{i \in h} d_{i,h_c},$$

---

[1]Strictly speaking, k-medoids minimizes the average distance to the representative center, but the sum is easier for computational reasons.

	1	2	3	4	5	6	7
1	0	3	4	7	8	11	10
2		0	3	6	7	10	9
3			0	3	4	7	6
4				0	1	4	3
5					0	3	2
6						0	3
7							0

Figure 7.3: Manhattan Inter-Point Distance Matrix

where h_c is a representative center for cluster h and d is the distance metric used (from a dissimilarity matrix). As was the case for K-Means (and K-Medians), the problem is NP hard and an exact solution does not exist.

The K-Medoids objective is identical in structure to that of a simple (unweighted) location-allocation facility location problem. Such problems consist of finding k optimal locations among n possibilities (the location part), and subsequently assigning the remaining observations to those centers (the allocation part).[2]

The main approach to solve the K-Medoids problem is the so-called *partitioning around medoids* (PAM) algorithm of Kaufman and Rousseeuw (2005). The logic underlying the PAM algorithm involves two stages, labeled *BUILD* and *SWAP*.

7.3.1 The PAM Algorithm for K-Medoids

In the *BUILD* stage, a set of k starting centers are selected from the n observations. In some implementations, this is a random selection, but Kaufman and Rousseeuw (2005), and, more recently Schubert and Rousseeuw (2019) prefer a step-wise procedure that optimizes the initial set.

The main part of the algorithm consists of the *SWAP* stage. This proceeds in a greedy iterative manner by swapping a current center with a candidate from the remaining noncenters, as long as the objective function can be improved.[3]

7.3.1.1 Build

The *BUILD* phase of the algorithm consists of identifying k observations out of the n and assigning them to be cluster centers h, with $h = 1, \ldots, k$. Kaufman and Rousseeuw (2005) outlines a step-wise approach that starts by picking the center, say h_1, that minimizes the overall sum of distances. This is readily accomplished by taking the observation that corresponds with the smallest row sum (or, equivalently, column sum) of the dissimilarity matrix.

The algorithm is again illustrated with the same toy example used in the previous two chapters. This consists of the seven-point coordinates listed in Figure 6.3, with $k = 2$ as the number of clusters. The corresponding Manhattan inter-point distance matrix is given in Figure 7.3. The row sums for each observation, are, respectively: 43, 32, 27, 24, 25, 37 and 33. The lowest value (24) is obtained for point 4, i.e., (6, 6), which is selected as the first cluster center.

---

[2]Location-allocation problems are typically solved by means of integer programming or specialized heuristics, see, e.g., Church and Murray (2009), Chapter 11.

[3]Detailed descriptions are given in Kaufman and Rousseeuw (2005), Chapters 2 and 3, as well as in Hastie et al. (2009), pp. 515–520, and Han et al. (2012), pp. 454–457.

	1	2	3	5	6	7	sum
	7	6	3	1	4	3	
1		3	0	0	0	0	3
2	4		0	0	0	0	4
3	3	3		0	0	0	6
5	0	0	0		1	1	2
6	0	0	0	0	4	0	4
7	0	5	0	0	1		6

Figure 7.4: BUILD – Step 2

ID	d_i4	d_i7
1	7	10
2	6	9
3	3	6
4	0	3
5	1	2
6	4	3
7	3	0

Figure 7.5: Cluster Assignment at end of BUILD stage

Next, at each step, an additional center (for $h = 2, \ldots, k$) is selected that improves the objective function the most. This is accomplished by evaluating each of the remaining points as a potential center. The one is selected that results in the greatest reduction in the objective function.

With the first center assigned, a $(n - 1) \times (n - 1)$ submatrix of the original distance matrix is used to assess each of the remaining $n - 1$ points (i.e., not including h_1) as a potential new center. This is implemented by computing the improvement that would result for each column point j from selecting a given row i as the new center. The row (destination) for which the improvement over all columns (origins) is the largest and is selected as the new center.

More specifically, in the second iteration the current distance between the point in column j and h_1 is compared to the distance from j to each row i. If j is closer to i than to its current closest center, then $d_{j,h_1} - d_{ji} > 0$, the potential improvement from assigning j to center i instead of to its current closest center. In the second iteration, this is h_1, but in later iterations it is more generally the closest center. Negative entries are not taken into account.

The row i for which the sum of the improvements over all j is the greatest becomes the next center. In the example, with point 4 as the first center, its distance from each of the remaining six points is listed in the second row of Figure 7.4, below the label of the point. Each of the following rows in the table gives the improvement in distance from the point at the top if the row point were selected as the new center, or $\max(d_{j4} - d_{ij}, 0)$. For example, for row 1 and column 2, that value is $6 - d_{1,2} = 6 - 3 = 3$. For the other points (columns), point 4 is closer than 2, so the corresponding matrix entries are zero, resulting in an overall row sum of 3 (the improvement in the objective from selecting 2 as the next center).

The row sum for each potential center is given in the column labeled *sum*. There is a tie between observations 3 and 7, which each achieve an improvement of 6. For consistency with the earlier chapters, point 7 is selected as the second starting point.

At this stage, the distance for each j to its nearest center is updated and the process starts anew for the $n - 2$ remaining observations. This process continues until all k centers have been assigned. This provides the initial solution for the *SWAP* stage. Since the example only has $k = 2$, the two initial centers are points 4 and 7.

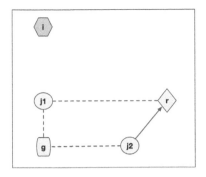

Figure 7.6: PAM SWAP Case 1

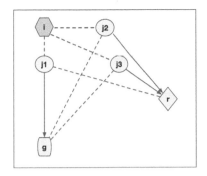

Figure 7.7: PAM SWAP Case 2

7.3.1.2 Swap

All the observations are assigned to the cluster center they are closest to. As shown in Figure 7.5, this results in cluster 1 consisting of observations 1-5, and cluster 2 being made up of observations 6 and 7, using the distances from columns 4 and 7 in the dissimilarity matrix. The total distance to each cluster center equals 20, with cluster 1 contributing 17 and cluster 2, 3

The essence of the *SWAP* phase is to consider what happens to the overall objective when a current center (one of the k) is swapped with one of the remaining $n - k$ noncenters. With a current center labeled as i and a noncenter as r, this requires the evaluation of $k \times (n - k)$ possible swaps (i, r).

When considering the effect of replacing a center i by a new center r, the impact on all the remaining $n - k - 1$ points must be considered. Those points can no longer be assigned to i, so they either become part of a cluster around the new center r, or they are re-allocated to one of the other centers g (but not i, since that is no longer a center).

For each point j, the change in distance associated with a swap (i, r) is labeled as C_{jir}. The total improvement for a given swap (i, r) is the sum of its effect over all j, $T_{ir} = \sum_j C_{jir}$. This total effect is evaluated for all possible (i, r) pairs and the minimum is selected. If that minimum is also negative, the associated swap is carried out, i.e., when $\operatorname{argmin}_{i,r} T_{ir} < 0$, r takes the place of i.

This process is continued until no more swaps can achieve an improvement in the objective function. The computational burden associated with this algorithm is quite high, since at each iteration $k \times (n - k)$ pairs need to be evaluated. On the other hand, no calculations other than comparison and addition/subtraction are involved, and all the information is in the (constant) dissimilarity matrix.

To compute the net change in the objective function for a point j that follows from a swap between i and r, two cases can be distinguished. In one, j was not assigned to center i, but it belongs to a different cluster, say g, such that $d_{jg} < d_{ji}$. In the other case, j does initially belong to the cluster i, such that $d_{ji} < d_{jg}$ for all other centers g. In both instances, the distances from j to the nearest current center (i or g) must be compared to the distance to the candidate point, r.

The first instance, where j is not part of the cluster i is illustrated in Figure 7.6. There are two scenarios for the configuration of the point j, labeled $j1$ and $j2$. Both these points are closer to g than to i, since they are *not* part of the cluster around i.

4,7 - 1	d_j4	d_j7	d_j1	C_j41	C_j71
2	6	9	3	-3	-3
3	3	6	4	1	0
5	1	2	8	1	0
6	4	3	11	0	1
T				-1	-2
4,7 - 2	d_j4	d_j7	d_j2	C_j42	C_j72
1	7	10	3	-4	-4
3	3	6	3	0	0
5	1	2	7	1	0
6	4	3	10	0	1
T				-3	-3
4,7 - 3	d_j4	d_j7	d_j3	C_j43	C_j73
1	7	10	4	-3	-3
2	6	9	3	-3	-3
5	1	2	4	1	0
6	4	3	7	0	1
T				-5	-5
4,7 - 5	d_j4	d_j7	d_j5	C_j45	C_j75
1	7	10	8	1	0
2	6	9	7	1	0
3	3	6	4	1	0
6	4	3	3	0	0
T				3	0
4,7 - 6	d_j4	d_j7	d_j6	C_j46	C_j76
1	7	10	11	3	0
2	6	9	10	3	0
3	3	6	7	3	0
5	1	2	3	1	0
T				10	0

Figure 7.8: PAM SWAP Cost Changes – Step 1

The key element is whether j is closer to r than to its current cluster center g. If $d_{jg} \leq d_{jr}$, then nothing changes, since j remains in cluster g, and $C_{jir} = 0$. This is the case for point $j1$. The dashed red line gives the distance to the current center g and the dashed green line gives the distance to r.

If j is closer to the new center r than to its current cluster g, then $d_{jr} < d_{jg}$, as is the case for point $j2$. As a result, j is assigned to the new center r and $C_{jir} = d_{jr} - d_{jg}$, a negative value, which decreases the overall cost. In the figure, the length of the dashed red line must be compared to the length of the solid green line, which designates a re-assignment of j to the candidate center r.

The second configuration is shown in Figure 7.7, where j starts out as being assigned to i. There are three possibilities for the location of j relative to g and r. In the first case, illustrated by point $j1$, j is closer to g than to r. This is illustrated by the difference in length between the dashed green line (d_{j1r}) and the solid green line (d_{j1g}). More precisely, $d_{jr} \geq d_{jg}$ so that j is now assigned to g. The change in the objective is $C_{jir} = d_{jg} - d_{ji}$. This value is positive, since j was part of cluster i and thus was closer to i than to g (compare the length of the red dashed line between $j1$ and i and the length of the line connecting $j1$ to g).

If j is closer to r, then $d_{jr} < d_{jg}$. There are two possible layouts, one depicted by $j2$, the other by $j3$. In both instances, the result is that j is assigned to r, but the effect on the objective differs.

In the Figure, for both $j2$ and $j3$ the dashed green line to g is longer than the solid green line to r. The change in the objective is the difference between the new distance and the old one (d_{ji}) or $C_{jir} = d_{jr} - d_{ji}$. This value could be either positive or negative, since what matters is that j is closer to r than to g, irrespective of how close j might have been to i. For point $j2$, the distance to i (dashed red line) was smaller than the new distance to r (solid green line), so $d_{jr} - d_{ji} > 0$. In the case of $j3$, the opposite holds, and the length to i

3,7 - 1	d_j3	d_j7	d_j1	C_j31	C_j71
2	3	9	3	0	0
4	3	3	7	0	0
5	4	2	8	0	2
6	7	3	11	0	4
T				0	6
3,7 - 2	d_j3	d_j7	d_j2	C_j32	C_j72
1	4	10	3	-1	-1
4	3	3	6	0	0
5	4	2	7	0	2
6	7	3	10	0	4
T				-1	5
3,7 - 4	d_j3	d_j7	d_j4	C_j34	C_j74
1	4	10	7	3	0
2	3	9	6	3	0
5	4	2	1	-1	-1
6	7	3	4	0	1
T				5	0
3,7 - 5	d_j3	d_j7	d_j5	C_j35	C_j75
1	4	10	8	4	0
2	3	9	7	4	0
4	3	3	1	-2	-2
6	7	3	3	0	0
T				6	-2
3,7 - 6	d_j3	d_j7	d_j6	C_j36	C_j76
1	4	10	11	6	0
2	3	9	10	6	0
4	3	3	4	0	0
5	4	2	3	0	1
T				12	1

Figure 7.9: PAM SWAP Cost Changes – Step 2

(dashed red line) is larger than the distance to the new center (solid green line). In this case, the change to the objective is $d_{jr} - d_{ji} < 0$.

In the example, the first step in the swap procedure consists of evaluating whether center 4 or center 7 can be replaced by any of the current noncenters, i.e., 1, 2, 3, 5 or 6. The comparison involves three distances for each point: the distance to the closest center, the distance to the second closest center and the distance to the candidate center. In the toy example, this is greatly simplified, since there are only two centers, with distances d_{j4} and d_{j7} for each noncandidate and noncenter point j. In addition, the distance to the candidate center d_{jr} is needed, where each noncenter point is in turn considered as a candidate (r).

All the evaluations for the first step are included in Figure 7.8. There are five main panels, one for each current noncenter point. The rows in each panel are the noncenter, noncandidate points. For example, in the top panel, 1 is considered a candidate, so the rows pertain to 2, 3, 5 and 6. Columns 3-4 give the distance to, respectively, center 4 (d_{j4}), center 7 (d_{j7}) and candidate center 1 (d_{j1}).

Columns 5 and 6 give the contribution of each row to the objective with point 1 replacing, respectively, 4 (C_{j41}) and 7 (C_{j71}). In the scenario of replacing point 4 by point 1, the distances from points 2 to 4 and 7 are 6 and 9, so point 2 is closest to center 4. As a result 2 will be allocated to either the new candidate center 1 or the current center 7. It is closest to the new center (distance 3 relative to 9). The decrease in the objective from assigning 2 to 1 rather than 4 is 3 - 6 = -3, the entry in the column C_{j41}.

Identical calculations are carried out to compute the contribution of 2 to the replacement of 7 by 1. Since 2 is closer to 4, this is the situation where a point is *not* closest to the center that is to be replaced. The assessment is whether point 2 would stay with its current center (4) or move to the candidate. Since 2 is closer to 1 than to 4, the gain from the swap is again -3, entered under C_{j71}. In the same way, the contributions are computed for each of

3,5 - 1	d_j3	d_j5	d_j1	C_j31	C_j51
2	3	7	3	0	0
4	3	1	7	0	2
6	7	3	11	0	4
7	6	2	10	0	4
T				0	10
3,5 - 2	d_j3	d_j5	d_j2	C_j32	C_j52
1	4	8	3	-1	-1
4	3	1	6	0	2
6	7	3	10	0	4
7	6	2	9	0	4
T				-1	9
3,5 - 4	d_j3	d_j5	d_j4	C_j34	C_j54
1	4	8	7	3	0
2	3	7	6	3	0
6	7	3	4	0	1
7	6	2	3	0	1
T				6	2
3,5 - 6	d_j3	d_j5	d_j6	C_j36	C_j56
1	4	8	11	4	0
2	3	7	10	4	0
4	3	1	4	0	2
7	6	2	3	0	1
T				8	3
3,5 - 7	d_j3	d_j5	d_j7	C_j37	C_j57
1	4	8	10	4	0
2	3	7	9	4	0
4	3	1	3	0	2
6	7	3	3	0	0
T				8	2

Figure 7.10: PAM SWAP Cost Changes – Step 3

the other noncenter and noncandidate points. The sum of the contributions is listed in the row labeled T. For a replacement of 4 by 1, the sum of -3, 1, 1 and 0 gives -1 as the value of T_{41}. Similarly, the value of T_{47} is the sum of -3, 0, 0 and 1, or -2.

The remaining panels show the results when each of the other current noncenters is evaluated as a center candidate. The minimum value over all pairs i, r is obtained for $T_{43} = -5$. This suggests that center 4 should be replaced by point 3 (there is actually a tie with T_{73}, so in each case 3 should enter the center set; in the example, it replaces 4). The improvement in the overall objective function from this step is -5.

This process is repeated in Figure 7.9, but now using d_{j3} and d_{j7} as reference distances. The smallest value for T is found for $T_{75} = -2$, which is also the improvement to the objective function (note that the improvement is smaller than for the first step, something that is to be expected from a gradient descent method). This suggests that 7 should be replaced by 5.

In the next step, shown in Figure 7.10, the calculations are repeated, using d_{j3} and d_{j5} as the distances. The smallest value for T is found for $T_{32} = -1$, suggesting that 3 should be replaced by 2. The improvement in the objective is -1 (again, smaller than in the previous step).

In the last step, shown in Figure 7.11, everything is recalculated for d_{j2} and d_{j5}. At this stage, none of the T yield a negative value, so the algorithm has reached a local optimum and stops.

The final result consists of a cluster of three elements, with observations 1 and 3 centered on 2, and a cluster of four elements, with observations 4, 6 and 7 centered on 5. As Figure 7.12 illustrates, both clusters contribute 6 to the total sum of deviations, for a final value of 12. This also turns out to be 20 - 5 - 2 - 1 or the total effect of each swap on the objective function.

2,5 - 1	d_j2	d_j5	d_j1	C_j21	C_j51
3	3	4	4	1	0
4	6	1	7	0	5
6	10	3	11	0	7
7	9	2	10	0	7
T				1	19
2,5 - 3	d_j2	d_j5	d_j3	C_j23	C_j53
1	3	8	4	1	0
4	6	1	3	0	2
6	10	3	7	0	4
7	9	2	6	0	4
T				1	10
2,5 - 4	d_j2	d_j5	d_j4	C_j24	C_j54
1	3	8	7	4	0
3	3	4	3	0	0
6	10	3	4	0	1
7	9	2	3	0	1
T				4	2
2,5 - 6	d_j2	d_j5	d_j6	C_j26	C_j56
1	3	8	11	5	0
3	3	4	7	1	0
4	6	1	4	0	3
7	9	2	3	0	1
T				6	4
2,5 - 7	d_j2	d_j5	d_j7	C_j27	C_j57
1	3	8	10	5	0
3	3	4	6	1	0
4	6	1	3	0	2
6	10	3	3	0	0
T				6	2

Figure 7.11: PAM SWAP Cost Changes – Step 4

ID	Cluster - 2	Cluster - 5
1	3	8
2	0	7
3	3	4
4	6	1
5	7	0
6	10	3
7	9	2
	6	6

Figure 7.12: PAM SWAP Final Assignment

7.3.2 Improving on the PAM Algorithm

The complexity of each iteration in the original PAM algorithm is of the order $k \times (n - k)^2$, which means it will not scale well to large data sets with potentially large values of k. To address this issue, several refinements have been proposed. The most familiar ones are CLARA, CLARANS and LAB.

7.3.2.1 CLARA

Kaufman and Rousseeuw (2005) proposed the algorithm CLARA, based on a sampling strategy. Instead of considering the full data set, a subsample is drawn. Then the PAM algorithm is applied to find the best k medoids in the sample. Finally, the distance from all observations (not just those in the sample) to their closest medoid is computed to assess the overall quality of the clustering.

The sampling process can be repeated for several more samples (keeping the best solution from the previous iteration as part of the sampled observations), and at the end the best solution is selected. While easy to implement, this approach does not guarantee that the best local optimum solution is found. In fact, if one of the best medoids is never sampled, it is impossible for it to become part of the final solution. Note that as the size of the sample becomes closer to the size of the full data set, the results will tend to be similar to those given by PAM.[4]

In practical applications, Kaufman and Rousseeuw (2005) suggest to use a sample size of 40 + 2k and to repeat the process 5 times.[5]

7.3.2.2 CLARANS

In Ng and Han (2002), a different sampling strategy is outlined that keeps the full set of observations under consideration. The problem is formulated as finding the best node in a graph that consists of all possible combinations of k observations that could serve as the k medoids. The nodes are connected by edges to the $k \times (n - k)$ nodes that differ in one medoid (i.e., for each edge, one of the k medoid nodes is swapped with one of the $n - k$ candidates).

The algorithm CLARANS starts an iteration by randomly picking a node (i.e., a set of k candidate medoids). Then, it randomly picks a neighbor of this node in the graph. This is a set of k medoids where one is swapped with the current set. If this leads to an improvement in the cost, then the new node becomes the new start of the next set of searches (still part of the same iteration). If not, another neighbor is picked and evaluated, up to *maxneighbor* times. This ends an iteration.

At the end of the iteration, the cost of the last solution is compared to the stored current *best*. If the new solution constitutes an improvement, it becomes the new *best*. This search process is carried out a total of *numlocal* iterations and at the end the best overall solution is kept. Because of the special nature of the graph, not that many steps are required to achieve a local minimum (technically, there are many paths that lead to the local minimum, even when starting at a random node).

Again, this can be illustrated by means of the toy example. To construct $k = 2$ clusters, any pair of 2 observations from the 7 could be considered a potential medoid. All those pairs constitute the nodes in the graph. The total number of nodes is given by the binomial coefficient $\binom{n}{k}$. In the example, $\binom{7}{2} = 21$.

Each of the 21 nodes in the graph has $k \times (n - k) = 2 \times 5 = 10$ *neighbors* that differ only in one medoid connected with an edge. For example, the initial node corresponding to (4,7) will be connected to all the nodes that differ by one medoid, i.e., either 4 or 7 is replaced (*swapped* in PAM terminology) by one of the $n - k = 5$ remaining nodes. Specifically, this includes the following 10 neighbors: 1-7, 2-7, 3-7, 5-7, 6-7, and 4-1, 4-2, 4-3, 4-5 and 4-6. Rather than evaluating all 10 potential swaps, as in PAM, only a maximum number (*maxneighbor*) are evaluated. At the end of those evaluations, the best solution is kept. Then the process is repeated, up to the specified total number of iterations, which Ng and Han (2002) call *numlocal*.

With *maxneighbors* set to 2, the first step of the random evaluation could be to *randomly* pick the pair 4-5 from the neighboring nodes. In other words, 7 in the original set is replaced by 5. Using the values from the worked example above, we have $T_{45} = 3$, a positive value, so this does not improve the objective function. The iteration count is increased (for *maxneighbors*) and a second random node is picked, e.g., 4-2 (i.e., replacing 7 by 2). At this stage, the

---

[4]Clearly, with a sample size of 100%, CLARA becomes the same as PAM.

[5]More precisely, the first sample consists of 40 + 2k random points. From the second sample on, the best k medoids found in a previous iteration are included, so that there are 40 + k additional random points. Also, in Schubert and Rousseeuw (2019), they suggested to use 80 + 4k and 10 repetitions for larger data sets. In the implementation in GeoDa, the latter is used for data sets larger than 100.

value $T_{42} = -3$, so the objective is improved to 20-3 = 17. Since the end of *maxneighbors* has been reached, this value is stored as *best* and the process is repeated with a different random starting point. This is continued until *numlocal* local optima have been obtained. At that point, the best overall solution is kept.

Based on several numerical experiments, Ng and Han (2002) suggest that no more than 2 iterations need to be pursued (i.e., *numlocal* = 2), with some evidence that more operations are not cost-effective. They also suggest a sample size of 1.25% of $k \times (n - k)$.[6]

Both CLARA and CLARANS are large data methods, since for smaller data sizes (e.g., $n < 100$), PAM will be feasible and obtain better solutions (since it implements an exhaustive evaluation).

Further speedup of PAM, CLARA and CLARANS is outlined in Schubert and Rousseeuw (2019), where some redundancies in the comparison of distances in the SWAP phase are removed. In essence, this exploits the fact that observations allocated to a medoid that will be swapped out, will move to either the second closest medoid or to the swap point. Observations that are not currently allocated to the medoid under consideration will either stay in their current cluster, or move to the swap point, depending on how the distance to their cluster center compares to the distance to the swap point. These ideas shorten the number of loops that need to be evaluated and allow the algorithms to scale to much larger problems (details are in Schubert and Rousseeuw, 2019, p.175). In addition, they provide an option to carry out the swaps for all current k medoids simultaneously, similar to the logic in K-Means, implemented in the FASTPAM2 algorithm (see Schubert and Rousseeuw, 2019, p.178).

7.3.2.3   LAB

A second type of improvement in the algorithm pertains to the BUILD phase. The original approach is replaced by a so-called *Linear Approximative BUILD* (LAB), which achieves linear runtime in n. Instead of considering all candidate points, only a subsample from the data is used, repeated k times (once for each medoid).

The FastPAM2 algorithm tends to yield the best cluster results relative to the other methods, in terms of the smallest sum of distances to the respective medoids. However, especially for large n and large k, FastCLARANS yields much smaller compute times, although the quality of the clusters is not as good as for FastPAM2. FastCLARA is always much slower than the other two. In terms of the initialization methods, LAB tends to be much faster than BUILD, especially for larger n and k.

7.3.3   Implementation

K-Medoids clustering is invoked from the drop-down list associated with the cluster toolbar icon, as the third item in the classic clustering methods subset, part of the full list shown in Figure 6.1. It can also be selected from the menu as **Clusters > K Medoids**.

This brings up the **KMedoids Clustering Settings** dialog. This dialog has an almost identical structure to the one for K-Medians, with a left-hand panel to select the variables and specify the cluster parameters, and a right-hand panel that lists the **Summary** results.

---

[6]Schubert and Rousseeuw (2019) also consider 2.5% in larger data sets with 4 iterations instead of 2.

```
Method: KMedoids (FastPAM)
Number of clusters: 8
Initialization method: LAB
Transformation: Standardize (Z)
Distance function: Manhattan
Medoids:
      275
      744
      567
      315
      608
      482
      335
      387
```

```
Cluster centers: (medoid)
   |<X-Centroids>| <Y-Centroids>|
--|-------------|--------------|
C1|1.16611e+06  |1.86192e+06
C2|1.15208e+06  |1.8878e+06
C3|1.16461e+06  |1.92994e+06
C4|1.15379e+06  |1.91443e+06
C5|1.1726e+06   |1.90617e+06
C6|1.18717e+06  |1.84695e+06
C7|1.1371e+06   |1.92202e+06
C8|1.1804e+06   |1.87197e+06
```

```
The total sum of squares:  1580
Within-cluster sum of squares:
   |Within cluster S.S.
--|--------------------
C1|34.6072
C2|43.083
C3|16.4323
C4|17.7079
C5|16.6174
C6|36.009
C7|30.1671
C8|10.1183
```

```
The total within-cluster sum of squares:   204.742
The between-cluster sum of squares:     1375.26
The ratio of between to total sum of squares:  0.870416
```

```
(Using Manhattan distance to medoids)
The total sum of distance: 1320.51
Within-cluster sum of distances:
   |Within Cluster D|Averages|
--|----------------|--------|
C1|71.2424         |0.624933
C2|77.9591         |0.702334
C3|49.8074         |0.456948
C4|48.6547         |0.463378
C5|45.9486         |0.483669
C6|66.272          |0.74463
C7|57.5422         |0.646542
C8|32.614          |0.412836
```

```
The total within-cluster sum of distance:  450.04
The ratio of total within to total sum of distance: 0.340808
```

Figure 7.13: Summary, K-Medoids, k=8

The method is illustrated with the *Chicago SDOH* sample data set. However, unlike the examples for K-Means and K-Medians, the variables selected are the X and Y coordinates of the census tracts. This showcases the location-allocation aspect of the K-Medoids method.

In the variable selection interface, the drop-down list contains **<X-Centroids>** and **<Y-Centroids>** as available variables, even when those coordinates have not been computed explicitly. They are included to facilitate the implementation of the spatially constrained clustering methods in Part III. The data layer is projected, with the coordinates expressed in feet, which makes for rather large numbers.

The default setup is to use the **Standardize (Z)** transformation, with the **FastPAM** algorithm using the **LAB** initialization. The distance function is set to **Manhattan**. As before, the number of clusters is chosen as **8**. There are no other options to be set, since the PAM algorithm proceeds with an exhaustive search in the SWAP phase, after the initial k medoids are selected.

Invoking **Run** populates the **Summary** panel with properties of the cluster solution and brings up a separate window with the cluster map. The cluster classification is saved as an integer variable with the specified **Field** name.

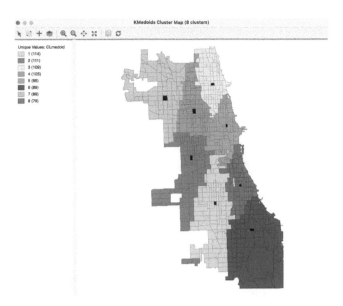

Figure 7.14: K-Medoids Cluster Map, k=8

7.3.3.1 Cluster results

The cluster characteristics are shown in Figure 7.13. Below the usual listing of methods and options, the *observation numbers* of the eight cluster medoids are given. These can be used to select the corresponding observations in the table. As before, the values for the different variables in each cluster center are listed, but now these correspond to actual observations. However, in the current application, the X and Y centroids are not part of the data table, unless they were added explicitly.

The customary sum of squares measures are included for completeness, but as pointed out in the discussion of K-Medians, these do not match the criteria used in the objective function. Nevertheless, the BSS/TSS ratio of 0.870416 shows a good separation between the eight clusters. Since the mean of the coordinates is close to the centroid, this ratio is actually not a totally inappropriate metric. However, in contrast to the objective function used, the distances are Euclidean and not Manhattan block distances.

The latter are employed in the calculation of the within-cluster sum of distances. As for K-Medians, both the total distance per cluster as well as the average are listed. The most compact cluster is C8, with an average distance of 0.413. The least compact cluster is C6, with an average of 0.745, closely followed by C2 (0.702). The ratio of within to total sum of the distances indicates a significant reduction to 0.34 of the original.

The spatial layout of the clusters is shown in the cluster map in Figure 7.14. In addition to the cluster membership for each tract, the cluster centers (medoid) are highlighted in black on the map. These are shown as tract polygons, although the actual cluster exercise was carried out for the tract centroids (points).[7]

The clusters show the optimal allocation of the 791 census tracts to 8 centers. The result is the most balanced of the cluster groupings obtained so far, with the four largest clusters consisting of between 114 and 105 tracts, and the four smallest between 95 and 79. In

---

[7]The map was created by first saving the selected cluster centers to a separate shape file and then using the GeoDa multilayer functionality to superimpose the centers on the cluster map.

addition, the clusters are almost perfectly spatially contiguous, even though K-Medoids is a nonspatial clustering method. This illustrates the application of standard clustering methods to location coordinates to delineate regions in space, a topic that is revisited in Chapter 9.

The one *outlier* tract that belongs to cluster 2 may seem to be a mistake, since it seems closer to the center of cluster 1 than to that of cluster 2, to which it has been assigned. However, the distances used in the algorithm are not straight line, but Manhattan distances. This case involves an instance where the centroid of the tract involved is almost directly south of the cluster 2 center, thus involving only a distance in the north-south direction. In the example, it turns out to be slightly smaller than the block distance required to reach the center of cluster 1.

7.3.4 Options and Sensitivity Analysis

The main option for K-Medoids is the choice of the **Method**. In addition to the default **FastPAM**, the algorithms **FastCLARA** and **FastCLARANS** are available as well. The latter two are large data methods. They will always perform (much) worse than FastPAM in small to medium-sized data sets.

In addition, there is a choice of **Initialization Method** between **LAB** and **BUILD**. In most circumstances, LAB is the preferred option, but BUILD is included for the sake of completeness and to allow for a full range of comparisons.

Each of these options has its own set of parameters.

7.3.4.1 CLARA parameters

The two main parameters that need to be specified for the CLARA method are the number of samples considered (by default set to 2) and the sample size. In addition, there is an option to include the best previous medoids in the sample.

7.3.4.2 CLARANS parameters

For CLARANS, the two relevant parameters pertain to the number of iterations and the sample rate. The former corresponds to the *numlocal* parameter in Ng and Han (2002), i.e., the number of times a local optimum is computed (default = 2). The sample rate pertains to the maximum number of *neighbors* that will be randomly sampled in each iteration (*maxneighbors* in the paper). This is expressed as a fraction of $k \times (n - k)$. The default is to use the value of 0.025 recommended by Schubert and Rousseeuw (2019), which would yield a maximum neighbors of 20 (0.025 x 791) for each iteration in the Chicago example. Unlike PAM and CLARA, there is no initialization option.

CLARANS is a large data method and is optimized for speed (especially with large n and large k). It should not be used in smaller samples, where the exhaustive search carried out by PAM can be computed in a reasonable time.

8

Spectral Clustering

This final chapter dealing with the treatment of classic clustering methods considers a totally different approach. **Spectral Clustering** is a graph partitioning method that can be interpreted as simultaneously implementing dimension reduction with cluster identification. It is designed to identify potentially nonconvex groupings in the multi-attribute space, something the other cluster methods are not able to do. It has become one of the go-to methods in modern machine learning.

Here again, the focus will be only on those aspects of spectral clustering that differ from what was covered in the previous chapters.

To illustrate spectral clustering, the well-known *spirals* data set from the literature will be used, shown in Figure 8.1. It is included as one of the `GeoDa` sample data sets. The distinctive pattern consists of 300 observations on two nonintersecting spiral point clouds.

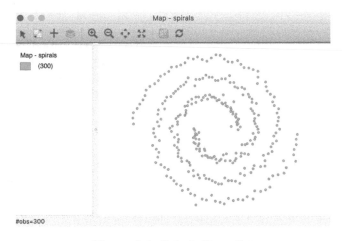

Figure 8.1: Spirals Data Set

8.1 Topics Covered

- Understand the graph-theoretic principles underlying spectral clustering
- Understand the importance of the graph Laplacian
- Carry out and interpret the results of spectral clustering
- Appreciate the sensitivity of cluster results to the parameters used

GeoDa Functions

- Clusters > Spectral

DOI: 10.1201/9781032713175-8

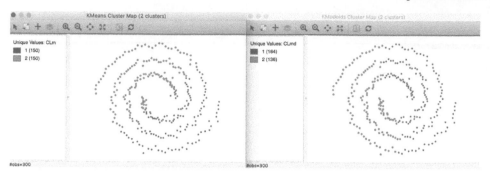

Figure 8.2: K-Means and K-Medoids for Spirals Data

8.2 Spectral Clustering Logic

Clustering methods like K-Means, K-Medians or K-Medoids are designed to discover convex clusters in the multidimensional data cloud. However, several interesting cluster shapes are not convex, such as the classic textbook spirals (Figure 8.1) or moons examples, or the famous *Swiss roll* and similar lower dimensional shapes embedded in a higher dimensional space.

These problems are characterized by a property that projections of the data points onto the original orthogonal coordinate axes (e.g., the X, Y, Z, etc. axes) do not create good separations. *Spectral clustering* approaches this issue by reprojecting the observations onto a new axes system and carrying out the clustering on the projected data points. Technically, this will boil down to the use of eigenvalues and eigenvectors, hence the designation as *spectral* clustering (see Section 2.2.2.1).

To illustrate the problem, consider the result of K-Means and K-Medoids (for $k = 2$) applied to the famous *spirals* data set from Figure 8.1. As Figure 8.2 shows, these methods tend to yield convex clusters, which fail to detect the nonlinear arrangement of the data. The K-Means clusters are arranged above and below a diagonal, whereas the K-Medoids result shows more of a left-right pattern.

In contrast, as shown in Figure 8.3, a spectral clustering algorithm applied to this data set perfectly extracts the two underlying patterns (initialized with a knn parameter of 3, see Section 8.4.3).

The mathematical solution underlying spectral clustering is based on a graph partitioning logic. More specifically, the clusters are viewed as subgraphs of a graph that represents the data structure. The subgraphs are internally well connected, but only weakly connected with the other subgraphs. Technical details can be found in the exhaustive overview of the various mathematical properties associated with spectral clustering contained in the *tutorial* by von Luxburg (2007).

In the remainder of the chapter, the graph partitioning idea is discussed first, followed by an overview of the main steps in a spectral clustering algorithm, including a review of some important parameters that need to be tuned in practice. The section closes with an illustration of its implementation, options and sensitivity analysis.

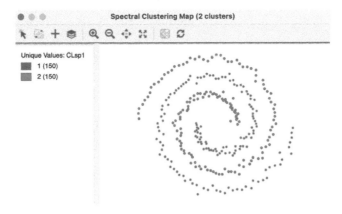

Figure 8.3: Spectral Clustering of Spirals Data

8.3 Clustering as a Graph Partitioning Problem

The clustering methods considered so far were based on the use of a *dissimilarity* matrix, with the objective of minimizing within-cluster dissimilarity. In spectral clustering, the focus is on the complement, i.e., a *similarity* matrix, and the goal is to maximize the internal similarity within a cluster. Of course, any dissimilarity matrix can be turned into a similarity matrix using a number of different methods. For example, a distance decay function can be employed (inverse distance, negative exponential) or a new metric can be computed as the difference from the maximum (e.g., $d_{max} - d_{ij}$).

A similarity matrix S consisting of elements s_{ij} can be viewed as the basis for the *adjacency matrix* of a weighted undirected graph $G = (V, E)$. In this graph, the vertices V are the observations and the edges E give the strength of the similarity between i and j, s_{ij}. This is identical to the interpretation of a spatial weights matrix (see Volume 1). As it turns out, the standard notation for the adjacency matrix is to use W, as was the case for spatial weights.

In practice, the adjacency matrix that is used in the algorithm is typically not the same as the full similarity matrix. Instead, it follows from a transformation of the latter to a *sparse* form (see Section 8.4.1).

The goal of graph partitioning is to delineate subgraphs that are internally strongly connected, but only weakly connected with the other subgraphs. In the ideal case, the subgraphs are so-called *connected components*. These are subgraphs with fully connected internal elements, but without any connections to the other subgraphs. A geographical example would be a collection of counties located on a number of islands. The counties on the same island are connected, but there are no connections between counties on different islands. As a result, the corresponding weight structure would be block-diagonal. In practice, this will rarely be the case.

The objective of graph partitioning then becomes one of finding a set of *cuts* in the graph that create k subsets, such that the internal connectivity is maximized and the in-between connectivity is minimized. A naive application of this principle would lead to the least connected vertices to become singletons (similar to what tends to happen in single and average linkage hierarchical clustering) and all the rest to form one large cluster. Better suited partitioning methods include a weighting of the cluster *size* so as to end up with well-balanced subset.[1]

A very important concept in this regard is the *graph Laplacian* associated with the adjacency matrix W.

8.3.1   Graph Laplacian

In network analysis or graph theory, an important property of the adjacency matrix W associated with a graph is the *degree* of a vertex i:

$$d_i = \sum_j w_{ij},$$

i.e., the row-sum of the similarity weights. This is the same concept as the cardinality of spatial weights, reflected in the connectivity histogram.

The *graph Laplacian* is defined as the following matrix:

$$L = D - W,$$

where D is a diagonal matrix containing the degree of each vertex. The graph Laplacian has the property that all its eigenvalues are real and nonnegative, and, most importantly, that its smallest eigenvalue is zero.

In the (ideal) case where the adjacency matrix can be organized into separate partitions (unconnected to each other), the matrix takes on a block-diagonal form. Each block contains the partitioning sub-matrix for the matching group. The corresponding graph Laplacian will similarly have a block-diagonal structure. Since each of these sub-blocks is itself a graph Laplacian (for the sub-network corresponding to the partition), its smallest eigenvalue is zero as well. An important result is then that if the graph is partitioned into k disconnected blocks, the graph Laplacian will have k zero eigenvalues. This forms the basis for the logic of using the k smallest eigenvalues of L to find the corresponding clusters.

While it can be used to proceed with spectral clustering, the unnormalized Laplacian L has some undesirable properties. Instead, the preferred approach is to use a so-called *normalized* Laplacian.

There are two ways to normalize the adjacency matrix and thus the associated Laplacian. One is to *row-standardize* the adjacency matrix or $D^{-1}W$. This is exactly the same idea as row-standardizing a spatial weights matrix. When applied to the Laplacian, this yields:

$$L_{rw} = D^{-1}D - D^{-1}W = I - D^{-1}W.$$

This is referred to as a *random walk* normalized graph Laplacian, since the row elements can be viewed as transition probabilities from state i to each of the other states j. As with spatial weights, the resulting normalized matrix is no longer symmetric, although its eigenvalues remain real, with the smallest eigenvalue being zero. The associated eigenvector is ι, a vector of ones.

A second transformation pre- and post-multiplies the Laplacian by $D^{-1/2}$, the inverse of the square root of the degree. This yields a *symmetric* normalized Laplacian as:

$$L_{sym} = D^{-1/2}DD^{-1/2} - D^{-1/2}WD^{-1/2} = I - D^{-1/2}WD^{-1/2}.$$

---

[1]For details, see Shi and Malik (2000), as well as the discussion of a "graph cut point of view" in von Luxburg (2007).

Again, the smallest eigenvalue is zero, but the associated eigenvector now becomes $D^{1/2}\iota$.

Spectral clustering algorithms differ by whether the unnormalized or normalized Laplacian is used to compute the k smallest eigenvalues and associated eigenvectors, and whether the Laplacian or the adjacency matrix is the basis for the calculation.

Specifically, as an alternative to using the smallest eigenvalues and associated eigenvectors of the normalized Laplacian, the *largest* eigenvalues/eigenvectors of the normalized adjacency (or affinity) matrix can be computed.

The standard eigenvalue expression is the following equality:

$$Lu = \lambda u,$$

where u is the eigenvector associated with eigenvalue λ. Subtracting both sides from Iu yields:

$$(I - L)u = (1 - \lambda)u.$$

In other words, if the smallest eigenvalue of L is 0, then the largest eigenvalue of $I - L$ is 1. Moreover:

$$I - L = I - [I - D^{-1/2}WD^{-1/2}] = D^{-1/2}WD^{-1/2}.$$

This result implies that the search for the smallest eigenvalue of L is equivalent to the search for the *largest* eigenvalue of $D^{-1/2}WD^{-1/2}$, the normalized adjacency matrix.

8.4 The Spectral Clustering Algorithm

In general terms, a spectral clustering algorithm consists of four phases:

- turning the similarity matrix into an adjacency matrix
- computing the k *smallest* eigenvalues and matching eigenvectors of the graph Laplacian, or, alternatively, the k *largest* eigenvalues and eigenvectors of the affinity matrix
- using the (rescaled) resulting eigenvectors to carry out K-Means clustering
- associating the resulting clusters back to the original observations

The most important step is the construction of the adjacency matrix.

8.4.1 Adjacency matrix

The first step consists of selecting a criterion to turn the *dense* similarity matrix into a *sparse* adjacency matrix, sometimes also referred to as the *affinity matrix*. The logic is very similar to that of creating spatial weights by means of a distance criterion.

For example, *neighbors* can be defined as being within a critical distance δ. In the spectral clustering literature, this is referred to as an epsilon (ϵ) criterion. This approach shares the same issues as in the spatial case when the observations are distributed with very different densities. In order to avoid isolates, a max-min nearest neighbor distance needs to be selected, which can result in a very unbalanced adjacency matrix. An adjacency matrix derived from the ϵ criterion is typically used in unweighted form.

A preferred approach is to use k nearest neighbors, although this is not a symmetric property. Since the eigenvalue computations require a symmetric matrix, there are two approaches to remedy the asymmetry (see also the discussion for spatial weights in Volume 1). In one, the affinity matrix is made symmetric as $(1/2)(W + W')$. In other words, if $w_{ij} = 1$, but $w_{ji} = 0$ (or the reverse), a new set of weights is created with $w_{ij} = w_{ji} = 1/2$.

Instead of a pure k-nearest neighbor criterion, so-called *mutual* k nearest neighbors can be defined, which consists of those neighbors among the k-nearest neighbor set that i and j have in common. More precisely, only those connectivities are kept for which $w_{ij} = w_{ji} = 1$.

The knn adjacency matrix can join points in disconnected parts of the graph, whereas the mutual k nearest neighbors will be sparser and tend to connect observations in regions with constant density. Each has pros and cons, depending on the underlying structure of the data.

An alternative approach is to compute a similarity matrix that has a built-in distance decay. The most common method is to use a Gaussian density (or *kernel*) applied to the Euclidean distance between observations:

$$s_{ij} = \exp[-(x_i - x_j)^2 / 2\sigma^2],$$

where the standard deviation σ plays the role of a bandwidth. With the right choice of σ, the corresponding adjacency matrix can be made more or less sparse. More precisely, the Gaussian transformation translates a dissimilarity matrix (Euclidean distances) into a similarity matrix. The new similarity matrix plays the role of the adjacency matrix in the remainder of the algorithm.

8.4.2   Clustering on the Eigenvectors of the Graph Laplacian

With the adjacency matrix in place, the k smallest eigenvalues and associated eigenvectors of the normalized graph Laplacian can be computed. Alternatively, the k largest eigenvalues/eigenvectors of the adjacency matrix can be calculated.

Since only a few eigenvalues are required, specialized algorithms can be exploited that only extract the smallest or largest eigenvalues/eigenvectors, such as the power iteration method outlined in Section 3.2.1.2.

When a symmetric normalized graph Laplacian is used, the $n \times k$ matrix of eigenvectors, say U, is row-standardized such that the norm of each row equals 1. The new matrix T has elements:[2]

$$t_{ij} = \frac{u_{ij}}{(\sum_{h=1}^{k} u_{ih}^2)^{1/2}}.$$

The new observations consist of the values of t_{ij} for each i. These values are used in a standard K-Means clustering algorithm to yield k clusters. Finally, the labels for the clusters are associated with the original observations and several cluster characteristics can be computed.

8.4.3   Spectral Clustering Parameters

The results of spectral clustering tend to be highly sensitive to the choice of the parameters that are used to define the adjacency matrix. For example, when using k-nearest neighbors, the choice of the number of neighbors is an important decision. In the literature, a value for the number of nearest neighbors of the order of $\log(n)$ is suggested for large n (von Luxburg, 2007). In practice, both $\ln(n)$ as well as $\log_{10}(n)$ are used.[3]

---

[2]Alternative normalizations are used as well. For example, sometimes the eigenvectors are rescaled by the inverse square root of the eigenvalues.

[3]In GeoDa, both options are available, with the values for the number of nearest neighbors *rounded up* to the nearest integer.

```
Number of clusters: 2
Affinity with K-Nearest Neighbors: K=3
Transformation: Standardize (Z)
Distance function: Euclidean
(K-Means) Initialization method:    KMeans++
(K-Means) Initialization re-runs:   150
(K-Means) Maximum iterations:   300

Cluster centers:
|   |x          |y         |
|---|-----------|----------|
|C1 | 0.019972  |-0.176148 |
|C2 |-0.0177098 | 0.177514 |

The total sum of squares:   598
Within-cluster sum of squares:
|   |Within cluster S.S. |
|---|--------------------|
|C1 | 285.729            |
|C2 | 287.864            |

The total within-cluster sum of squares:   573.594
The between-cluster sum of squares:     24.4061
The ratio of between to total sum of squares:  0.0408129
```

Figure 8.4: Cluster Characteristics for Spectral Clustering of Spirals Data – Default Settings

Similarly, the bandwidth of the Gaussian transformation is determined by the value for the standard deviation, σ. One suggestion for the value of σ is to take the mean distance to the k nearest neighbor or $\sigma \sim \log(n) + 1$ (von Luxburg, 2007). Again, either $\ln(n) + 1$ or $\log_{10}(n) + 1$ can be implemented. In addition, sometimes $\sigma = \sqrt{1/p}$ is suggested, where p corresponds to the number of variables (features).

In practice, these parameters are best set by trial and error, and a careful sensitivity analysis is in order.

8.5 Implementation

Spectral clustering is invoked from the **Clusters** toolbar, as the next to last item in the classic clustering subset, shown in Figure 6.1. Alternatively, from the menu, it is selected as **Clusters > Spectral**.

The variable settings panel has the same general layout as for K-Means. The coordinates for the points are selected from the **Select Variables** dialog as **x** and **y**. In the **Parameters** panel, one set pertains to the construction of the **Affinity** (or adjacency) matrix, the others are the usual parameters for the K-Means algorithm that is applied to the transformed eigenvectors.

The **Affinity** option provides three alternatives, **K-NN**, **Mutual K-NN** and **Gaussian**, with specific parameters for each. In the example, the number of clusters is set to 2 and all options are kept to the default setting. This includes **K-NN** with 3 neighbors for the affinity matrix and all the default settings for K-Means. The value of 3 for knn corresponds to $\log_{10}(300) = 2.48$, rounded up to the next integer.

The **Run** button generates the cluster characteristics in the **Summary** panel, as in Figure 8.4. In the usual manner, this also brings up a new window with the cluster map and saves the cluster classification as an integer variable to the data table.

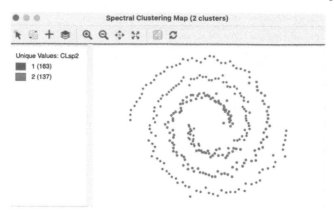

Figure 8.5: Cluster Characteristics for Spectral Clustering of Spirals Data – Mutual KNN=3

```
Number of clusters: 2
Affinity with Mutual K-Nearest Neighbors:  K=3
Transformation: Standardize (Z)
Distance function: Euclidean
(K-Means) Initialization method:    KMeans++
(K-Means) Initialization re-runs:   150
(K-Means) Maximum iterations:  300

Cluster centers:
   |x          |y
---|-----------|---------
C1 |-0.0950711 |0.0960661
C2 |0.115591   |-0.112803

The total sum of squares:  598
Within-cluster sum of squares:
   |Within cluster S.S.
---|--------------------
C1 |410.848
C2 |170.341

The total within-cluster sum of squares:  581.189
The between-cluster sum of squares:    16.8112
The ratio of between to total sum of squares: 0.0281123
```

Figure 8.6: Cluster Characteristics for Spectral Clustering of Spirals Data – Mutual KNN=3

8.5.1   Cluster results

The cluster map for the chosen parameter settings is shown in Figure 8.3. It yields a perfect separation of the two spirals, although that is by no means the case for all parameter settings, as will be illustrated below.

The cluster characteristics in Figure 8.4 list the parameter settings first, followed by the values for the cluster centers (the mean) for the two (standardized) coordinates and the decomposition of the sum of squares. The ratio of between sum of squares to total sum of squares is a dismal 0.04. This is not surprising, since this criterion provides a measure of the degree of compactness for the cluster, which a nonconvex cluster like the spirals example does not meet.

In this example, it is easy to visually assess the extent to which the nonlinearity is captured. However, in the typical high-dimensional application, this will be much more of a challenge, since the usual measures of compactness may not be informative. A careful inspection of the distribution of the different variables across the observations in each cluster is therefore in order.

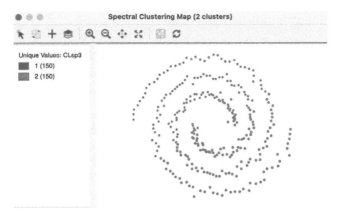

Figure 8.7: Cluster Characteristics for Spectral Clustering of Spirals Data – Gaussian Kernel

```
Number of clusters: 2
Affinity with Gaussian Kernel: Sigma=0.707107
Transformation: Standardize (Z)
Distance function:  Euclidean
(K-Means) Initialization method:    KMeans++
(K-Means) Initialization re-runs:   150
(K-Means) Maximum iterations:   300

Cluster centers:
   |  x        |  y        |
 --|-----------|-----------|
 C1|-0.276703  | 0.475971  |
 C2| 0.278965  |-0.474606  |

The total sum of squares:   598
Within-cluster sum of squares:
   |  Within cluster S.S.|
 --|---------------------|
 C1| 182.494             |
 C2| 181.874             |

The total within-cluster sum of squares:   364.367
The between-cluster sum of squares:    233.633
The ratio of between to total sum of squares:  0.39069
```

Figure 8.8: Cluster Characteristics for Spectral Clustering of Spirals Data – Gaussian Kernel

8.5.2 Options and Sensitivity Analysis

The results of spectral clustering are extremely sensitive to the parameters chosen to create the affinity matrix. Suggestions for default values are only suggestions, and the particular values may sometimes be totally unsuitable. Experimentation is therefore a necessity. There are two classes of parameters. One set pertains to the number of nearest **Neighbors** for **K-NN** or **Mutual K-NN**. The other set relates to the bandwidth of the **Gaussian** kernel, determined by the standard deviation **Sigma**.

8.5.2.1 K-nearest neighbor affinity matrix

The two default values for the number of nearest neighbors are contained in a drop-down list. In the spirals example, with n=300, $\log_{10}(n) = 2.48$, which rounds up to 3, and $\ln(n) = 5.70$, which rounds up to 6. These are the two default values provided. Any other value can be entered manually in the dialog. The options for a mutual knn affinity matrix have the same entries.

The results for the **Mutual** option with 3 nearest neighbors are shown in Figures 8.5 and 8.6. The separation is far from perfect, with cluster members in both spirals. The measures of fit are even worse than for the default case.

8.5.2.2   Gaussian kernel affinity matrix

The built-in options for **sigma**, the standard deviation of the **Gaussian** kernel are 0.707107, 3.477121 and 6.703782. The smallest value corresponds to $\sqrt{1/p}$, where p, the number of variables, equals 2 in the example. The other two values are $\log_{10}(n) + 1$ and $\ln(n) + 1$, yielding respectively 3.477121 and 6.703782 for n=300. In addition, any other value can be entered in the dialog.

The results for the first option (0.707107) are shown in Figures 8.7 and 8.8. The clusters totally fail to extract the shape of the separate spirals, although they are perfectly balanced. The layout looks similar to the results for K-Means and K-Medians in Figure 8.2. Interestingly, this layout scores much better on the BSS/TSS ratio (0.39). Rather than being an indication of a good separation, it suggests that the nonlinearity of the true clusters is not reflected in the grouping.

In order to find a solution that provides the same separation as in Figure 8.3, some experimentation with different values for σ is needed. As it turns out, the same result as for knn with 3 neighbors is obtained with $\sigma = 0.08$ or 0.07, neither of which are even close to the default values. This illustrates how in an actual example, where the results cannot be readily visualized in two dimensions, it may be very difficult to find the parameter values that discover the true underlying patterns.

Part III

Spatial Clustering

9

Spatializing Classic Clustering Methods

In Part III, the focus shifts to how spatial aspects of the data can be included explicitly into cluster analysis. Foremost among these aspects are location and contiguity, as a way to formalize *locational similarity*. The treatment in this chapter is largely pedagogical, aimed at illustrating the tension between attribute similarity and locational similarity. This is carried out through methods that *spatialize* classic cluster techniques. The clustering methods are the same as covered in the preceding chapters, i.e., Hierarchical Clustering, K-Means, K-Medians, K-Medoids, and Spectral Clustering. They all are based on nonspatial considerations to group the data. In this chapter, the spatial dimension is introduced in four different ways.

First, classic methods can be applied to geographical coordinates to create *regions* that are purely based on location in geographical space. This does not consider other attributes. An illustration of this approach was already given in the discussion of K-Medoids, in Chapter 7.

The next two sets of methods attempt to construct a compromise between attribute similarity and locational similarity, with different degrees of forcing a *hard* contiguity constraint. Both approaches use standard clustering techniques. In one, the feature set (i.e., the variables under consideration) is expanded with the coordinates of the location. This provides some weight to the locational pull (spatial compactness) of the observations, although this is by no means binding. In the other approach, the problem is turned into a form of multi-objective optimization, where both the objective of attribute similarity and the objective of geographical similarity (co-location) are weighted so that a compromise between the two can be evaluated.

A final method aims to turn the nonspatial solution of a classic clustering method into a set of spatially contiguous groupings, i.e., where the members of each cluster are also spatially connected. The heuristic outlined is primarily intended to illustrate the trade-offs between attribute and locational similarity. It is usually less than optimal compared to the spatially explicit methods covered in the next two chapters.

The spatial clustering methods are illustrated by means of the *Ceará Zika* sample set with 184 municipal entities. The specific variables are the five urban dimensions included in the Brazilian index of urban structure (IBEU): mobility, environmental conditions, housing conditions, sanitation and infrastructure. The indices are composites of other variables and are scaled between zero and one (larger values indicating better conditions). Following the analysis in Amaral et al. (2019), GDP per capita is included as well. These indicators are illustrative of the type of variables typically used in a regional clustering exercise.

DOI: 10.1201/9781032713175-9

9.1   Topics Covered

- Understand the difference between spatial and a-spatial clustering
- Apply standard clustering methods to geographic coordinates
- Assess the effect of including geographical coordinates as cluster variables
- Gain insight into the trade-off between attribute similarity and contiguity in classic clustering methods

GeoDa Functions

- Clusters > K Means
- Clusters > K Medoids
- Clusters > Spectral
- Clusters > Hierarchical
    - use geometric centroids
    - auto weighting
- Clusters > Make Spatial
- Clusters > SC K Means

9.2   Clustering on Geographic Coordinates

Applying classic cluster methods to geographical coordinates results in clusters as *regions* in space. There is nothing special to this type of application. In fact, a number of examples of this approach have already been illustrated in previous chapters.

When using actual geographical coordinates as the features in a clustering exercise, it is important to make sure that the **Transformation** is set to **Raw**. This is to avoid distortions in the distance measure that may result from the transformations, such as z-standardize.

For example, in cases where one dimension dominates the other, the range of the coordinates in one direction (e.g., North-South) can be much greater than the range in the other direction (East-West). Standardization of the coordinates would result in the transformed values to be more *compressed* in the dominant direction, since both transformed coordinates end up with the same variance. This will directly affect the resulting inter-point distances used in the dissimilarity matrix.

In addition, it is critical that the geographical coordinates are *projected* and not simply latitude-longitude decimal degrees. This ensures that the Euclidean distances are calculated correctly. Using latitude and longitude to calculate straight line distance is incorrect.

For most methods, the clusters will tend to result in fairly compact regions. For example, for K-Means, the regions correspond to Thiessen polygons around the cluster centers. K-Medoids are the solution to a form of location-allocation problem, with the cluster center serving as the *location* and the cluster members containing the *allocation* or service area. For spectral clustering, the boundaries between regions can be more irregular. Finally, for hierarchical clustering, the result depends greatly on the linkage method chosen. In most situations, Ward's and complete linkage will yield the most compact regions. In contrast, single and average linkage will tend to result in singletons and/or long chains of points.

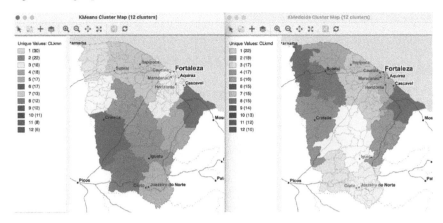

Figure 9.1: K-Means and K-Medoids on X-Y Coordinates

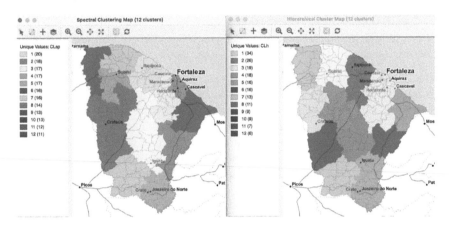

Figure 9.2: Spectral and Hierarchical on X-Y Coordinates

9.2.1 Implementation

The cluster methods are invoked in the standard manner, as illustrated in Chapters 5 to 8. The only difference is in the selection of variables.

If included among the data set variables, X and Y coordinates can be selected in the same way as any other variable. This is quite flexible and allows any pair of variables to be chosen as *coordinates*, although to some extent, that defeats the purpose.

An alternative is to select the coordinates as **<X-Centroids>** and **<Y-Centroids>** (at the bottom of the variables list in the Clustering Settings dialog), as in Section 7.3.3. This selection invokes a check on the proper projection (a warning will be generated for latitude-longitude decimal degrees). As mentioned, the **Transformation** is best set to **Raw** to keep the original dimensions.

The corresponding cluster maps are shown in Figures 9.1 and 9.2, respectively for K-Means and K-Medoids, and for Spectral and Hierarchical clustering. The number of clusters is set to **12**. In each instance, the defaults were used, with knn=8 for Spectral Clustering and Ward's linkage for Hierarchical Clustering.[1] The clusters vary slightly in fit, with K-Means obtaining the highest BSS/TSS ratio at 0.941, followed closely by Hierarchical Clustering

---

[1]The maps are shown with a Stamen > Toner Lite base layer in order to situate the clusters in context.

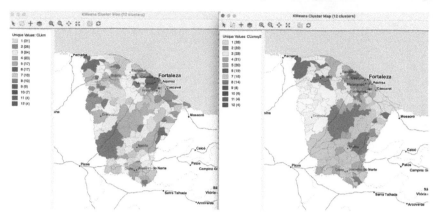

Figure 9.3: K-Means without and with X-Y Coordinates included

(0.937) and K-Medoids (0.933). Spectral Clustering does slightly less well at 0.929. Overall, however, these results reveal excellent cluster separation.

Nevertheless, the corresponding spatial patterns are quite distinct. The largest clusters range from 34 observations for Hierarchical Clustering to 20 for Spectral Clustering. For all but K-Medoids, the location of the largest cluster is in the North-East of the state, but for K-Medoids, it is at the South-Eastern tip of the state. The spatial layout is similar in broad strokes, with most clusters fairly compact and aligning with the state border. One to two clusters make up the central regions. However, there are many differences as well. This illustrates that even when only spatial considerations are taken into account, the respective methods result in varying layouts.

More formal comparisons of the spatial features of clusters are considered in Chapter 12.

9.3   Including Geographical Coordinates in the Feature Set

Classic clustering procedures only deal with attribute similarity. They therefore do not guarantee that the resulting *clusters* are spatially contiguous, nor are they designed to do so. A number of ad hoc procedures have been suggested that start with the original result and proceed to create groupings where all members of the cluster become contiguous. In general, such approaches are unsatisfactory and none scale well to larger data sets. However, they are useful to illustrate the trade-offs involved in going from a nonspatial to a spatial solution.

Early approaches started with the solution of a standard clustering algorithm, and subsequently adjusted this in order to obtain contiguous regions, as in Openshaw (1973) and Openshaw and Rao (1995). For example, one could make every subset of contiguous observations a separate *cluster*, which would satisfy the spatial constraint. However, this would also increase the number of clusters. As a result, the initial value of k would no longer be valid. One could also *manually* move observations to an adjoining cluster and visually obtain contiguity while maintaining the same k. However, this quickly becomes impractical when the number of observations is larger. A simple and scalable heuristic that implements this idea is illustrated in Section 9.5.

```
Method: KMeans
Number of clusters: 12
Initialization method: KMeans++
Initialization re-runs: 150
Maximum iterations: 1000
Transformation: Standardize (Z)
Distance function: Euclidean

Cluster centers:
    |mobility|environ |housing |sanitation|infra   |gdpcap  |
|---|--------|--------|--------|----------|--------|--------|
C1  |0.973677|0.912677|0.854323|0.617     |0.457452|4.68184 |
C2  |0.96748 |0.90492 |0.8378  |0.80112   |0.54308 |5.1908  |
C3  |0.95175 |0.88375 |0.803625|0.648875  |0.555542|4.50504 |
C4  |0.952087|0.747304|0.809565|0.576739  |0.496565|4.19865 |
C5  |0.956176|0.887118|0.780059|0.516824  |0.439059|5.361   |
C6  |0.974353|0.942765|0.860765|0.692059  |0.613647|4.19265 |
C7  |0.948692|0.825692|0.858692|0.623538  |0.564   |4.80246 |
C8  |0.9655  |0.6647  |0.8526  |0.6768    |0.6236  |5.675   |
C9  |0.919   |0.91    |0.848111|0.552778  |0.398889|5.15233 |
C10 |0.911429|0.707286|0.733571|0.607714  |0.438714|4.71886 |
C11 |0.9525  |0.85925 |0.81225 |0.56375   |0.43125 |27.7873 |
C12 |0.833   |0.76575 |0.818   |0.7925    |0.57375 |12.6005 |

The total sum of squares:  1098
Within-cluster sum of squares:
    |Within cluster S.S.|
|---|-------------------|
C1  |52.1157            |
C2  |42.0426            |
C3  |37.9508            |
C4  |35.9542            |
C5  |33.5191            |
C6  |26.218             |
C7  |15.2013            |
C8  |27.4693            |
C9  |18.2422            |
C10 |27.0935            |
C11 |22.227             |
C12 |11.0881            |

The total within-cluster sum of squares:   349.122
The between-cluster sum of squares:    748.878
The ratio of between to total sum of squares:  0.682038
```

Figure 9.4: K-Means Cluster Characteristics

An alternative approach is to include the geometric centroids of the areal units as part of the clustering process by adding them as variables in the collection of attributes. Early applications of this idea can be found in Webster and Burrough (1972) and Murray and Shyy (2000), among others. However, while this tends to yield more geographically compact clusters, it does not guarantee contiguity. In other words, it introduces a *soft* spatial constraint.

In this context, it is again important to ensure that the coordinates are in projected units. Even though it is sometimes suggested in the literature to include latitude and longitude as additional features, this is incorrect. Only for projected units are the associated Euclidean distances meaningful.

Also, as pointed out earlier, the transformation of the variables, which is standard in the cluster procedures, may distort the geographical features when the East-West and North-South dimensions are very different. Due to the standardization, smaller distances in the shorter dimension will have the same weight as larger distances in the longer dimension, since the latter are more compressed in the standardization.

Unlike what was the case when clustering solely on geographic coordinates, keeping all the variables (including the nongeographical attributes) in a **Raw** format is *not advisable*. Due to the standardization, the result will be somewhat biased in favor of compactness in the longer dimension.

9.3.1 Implementation

To illustrate the inclusion of coordinates among the features, a K-Means analysis is carried out for the five urban dimension indices and the GDP per capita for the 184 municipalities in Ceará. The corresponding cluster map is shown in the left-hand panel of Figure 9.3, with the cluster characteristics listed in Figure 9.4.

To set the baseline without coordinates, the six variables **mobility**, **environ**, **housing**, **sanitation**, **infra** and **gdpcap** are specified in the **KMeans Clustering Settings** dialog. All the default parameter values are used, with the number of clusters set to **12**.

```
Method: KMeans
Number of clusters: 12
Initialization method: KMeans++
Initialization re-runs: 150
Maximum iterations: 1000
Transformation: Standardize (Z)
Distance function: Euclidean

Cluster centers:
    |mobility|environ |housing |sanitation|infra   |gdpcap  |<X-Centroids>|<Y-Centroids>|
 —|————————|————————|————————|—————————|————————|—————————|——————————|——————————|
C1  |0.962962|0.885462|0.833846|0.690692  |0.569692|4.55896 |5.29118e+06 |9.56658e+06  |
C2  |0.946913|0.860261|0.785087|0.536     |0.478696|4.79848 |5.42019e+06 |9.56949e+06  |
C3  |0.962043|0.892565|0.84987 |0.59      |0.481304|4.42665 |5.26664e+06 |9.52517e+06  |
C4  |0.966524|0.909286|0.847714|0.682619  |0.594476|4.17338 |5.40175e+06 |9.28994e+06  |
C5  |0.954   |0.70385 |0.84435 |0.6361    |0.5566  |4.9721  |5.41323e+06 |9.26669e+06  |
C6  |0.966   |0.903947|0.843158|0.804632  |0.548947|5.36432 |5.42256e+06 |9.42422e+06  |
C7  |0.951563|0.906125|0.845125|0.593813  |0.4165  |6.18844 |5.53651e+06 |9.47832e+06  |
C8  |0.975071|0.902929|0.823786|0.609286  |0.445357|4.164   |5.41165e+06 |9.24936e+06  |
C9  |0.90975 |0.72825 |0.753875|0.647125  |0.47475 |4.79375 |5.46627e+06 |9.52835e+06  |
C10 |0.956833|0.715833|0.768   |0.539667  |0.500333|4.20283 |5.23629e+06 |9.63866e+06  |
C11 |0.833   |0.76575 |0.818   |0.7925    |0.57375 |12.6005 |5.485e+06   |9.57432e+06  |
C12 |0.9525  |0.85925 |0.81225 |0.56375   |0.43125 |27.7873 |5.50493e+06 |9.53911e+06  |

The total sum of squares:  1464
Within-cluster sum of squares:
    |Within cluster S.S.|
 —|————————————————|
C1  |85.7498
C2  |58.0514
C3  |56.2203
C4  |44.3279
C5  |74.0855
C6  |56.9298
C7  |45.8394
C8  |51.2818
C9  |37.0142
C10 |14.3513
C11 |11.1522
C12 |23.814

The total within-cluster sum of squares:   558.818
The between-cluster sum of squares:    905.182
The ratio of between to total sum of squares:  0.618294
```

Figure 9.5: K-Means Cluster Characteristics with X-Y Coordinates

The resulting spatial pattern is quite disparate, with cluster sizes ranging from 31 to 4. The only spatially compact cluster is the smallest, C12, consisting of 4 municipalities concentrated around the largest city of Fortaleza. In contrast, C11, also consisting of 4 municipalities, is made up of 4 completely spatially separated entities. For the other clusters, the ratio of subclusters to total cluster members, an indicator of fragmentation (higher values indicating more fragmentation), ranges from 0.38 for C3 (9 subclusters for 24 members) to 0.70 for C8 (7 subclusters for 10 members) and 0.71 for C5 and C6 (12 subclusters for 17 members). On the other hand, the overall separation is quite good, with a BSS/TSS ratio of 0.682.

As always, a careful interpretation of the cluster characteristics can be derived from the cluster center values for each of the variables, but this is beyond the current scope.

The coordinates are included among the features by adding **<X-Centroids>** and **<Y-Centroids>** to the six variables already specified in the **KMeans Clustering Settings** dialog. The rest of the interface remains unchanged.

The resulting cluster map is included as the right-hand panel in Figure 9.3, with the cluster characteristics in Figure 9.5. The clusters are more spatially compact, although by no means contiguous. C8 still consists of 9 subclusters (out of 14 members), and C1 and C2 consist of 8 (respectively, out of 26 and 23 cluster members). On the other hand, C10 only has two subclusters, and C3, C4, C7 and C9 only have 3. The BSS/TSS ratio decreases to 0.618, but it is not directly comparable, since it also includes the X and Y coordinates in the computations.

Since the X and Y coordinates were included in the analysis as regular variables, they are also included in the statistics for cluster centers in Figure 9.5.

The results illustrate that including the coordinates may provide a form of spatial forcing, albeit imperfect. The outcome will tend to vary by clustering technique and the number of clusters specified (k), as well as by the number of attribute variables. In essence, this method gives equal weight to all the variables, so the higher the attribute dimension of the input variables, the less weight the spatial coordinates will receive. As a consequence, less spatial forcing will be possible with more variables. The heuristic outlined in the next section addresses this issue.

9.4 Weighted Optimization of Geographical and Attribute Similarity

In the previous section, each variable was weighted equally when including geographic coordinates among the features in the clustering exercise. In contrast, in a weighted optimization, the coordinate variables are treated separately from the regular attributes, in the sense that the problem is now formulated as having *two objective functions*. One objective is focused on the similarity of the regular attributes (e.g., the six urban indicator variables), and the other on the similarity of the geometric centroids. A weight changes the relative importance of each objective.

Early formulations of the idea behind a weighted optimization of geographical and attribute features can be found in Webster and Burrough (1972), Wise et al. (1997) and Haining et al. (2000). More recently, some other methods that have a *soft* enforcing of spatial constraints can be found in Yuan et al. (2015) and Cheruvelil et al. (2017), applied to spectral clustering, and in Chavent et al. (2018) for hierarchical clustering. In these approaches, a relative weight is given to the dissimilarity that pertains to the regular attributes and one that pertains to the geographic coordinates. As the weight of the geographic part is increased, more spatial forcing occurs.

The approach outlined here is different, in that the weights are used to rescale the original variables in order to change the relative importance of attribute similarity versus locational similarity. This is a special case of *weighted* clustering, which is an option in several of the standard clustering implementations. The objective is to find a contiguous solution with the *smallest* weight given to the X-Y coordinates, thus sacrificing the least with respect to attribute similarity.

Consider there to be p regular attributes in addition to the X-Y coordinates. With a weight of $w_1 \leq 1$ for the geographic coordinates, the weight for the regular attributes as a whole is $w_2 = 1 - w_1$. These weights are distributed over the variables to achieve a rescaling that reflects the relative importance of space vs. attribute.

Specifically, the weight w_1 is allocated to X and Y with each $w_1/2$, whereas the weight w_2 is allocated over the regular attributes as w_2/p. For comparison, and using the same logic, in the method outlined in the previous section, each variable was given an equal weight of $1/(p+2)$. In other words, all the variables, both attribute and locational were given the same weight.

The rescaled variables can be used in all of the standard clustering algorithms, without any further adjustments. One other important difference with the method in the previous section is that the X-Y coordinates are *not* taken into account to compute the cluster centers and assess performance measures. Those are based on the original unweighted attributes.

While the principle behind this approach is intuitive, it does not always work in practice, since the contiguity constraint is not actually part of the optimization process. Therefore, there are situations where simple re-weighting of the spatial and nonspatial attributes does not lead to a contiguous solution.

For all methods except spectral clustering, there is a direct relation between the relative weight given to the attributes and the measure of fit (e.g., the between to total sum of squares). As the weight for the X-Y coordinates increases, the fit on the attribute dimension will become poorer. However, for spectral clustering, this is not necessarily the case, since this method is not based on the original variables, but on a projection of these (the principal components).

In other words, the X-Y coordinates are used to pull the original unconstrained solution toward a point where all clusters consist of contiguous observations. Only when the weight given to the coordinates is *small* is such a solution still meaningful in terms of the original attributes. A large weight for the X-Y coordinates in essence forces a contiguous solution, similar to what would follow if only the coordinates were taken into account. While this obtains contiguity, it typically does not provide a meaningful result in terms of attribute similarity.

9.4.1   Optimization

The spatial constraint can be incorporated into an optimization process. In the heuristic outlined here, a bisection search is employed to find the cluster solution with the *smallest* weight for the X-Y coordinates that still satisfies a contiguity constraint. In most instances, this solution will also have the best fit of all the possible weights that satisfy the constraint. For spectral clustering, the solution only guarantees that it consists of contiguous clusters with the largest weight given to the original attributes.

The heuristic contains two important steps: changing the weight at each iteration in an efficient manner, and assessing whether the solution consists of contiguous clusters.

A limitation of this approach is that it does not deal well with isolates (islands), i.e., disconnected observations. Consequently, only results that designate the isolate(s) as separate clusters will meet the contiguity constraint. In such instances, it may be more practical to remove the isolate observations before carrying out the analysis.

The method outlined in the previous section implements *soft* spatial clustering, in the sense that not all clusters consist of spatially contiguous observations. In contrast, the approach outlined here enforces a *hard* spatial constraint, although possibly at the expense of a poor clustering performance for the attribute variables.

The point of departure is a purely spatial cluster that assigns a weight of $w_1 = 1.0$ to the X-Y coordinates. This yields a set of spatially compact clusters for the given value of k. Next, w_1 is set to 0.5 and the contiguity constraint is checked. As is customary, contiguity is defined by a spatial weights specification. The spatial weights are represented internally as a graph. For each node in the graph, the algorithm keeps track of what cluster it belongs to.

If the contiguity constraint is satisfied for $w_1 = 0.5$, that means that w_1 can be further decreased, to get a better fit on the attribute clustering. Using a bisection logic, the weight is set to 0.25 and the contiguity constraint is checked again.

In contrast, if the contiguity constraint is not satisfied at $w_1 = 0.5$, then the weight is increased to the mid-point between 0.5 and 1.0, i.e., 0.75, and the process is repeated. Each time, a failure to meet the contiguity constraint increases the weight, and the reverse results in a decrease of the weight. Following the bisection logic, the next point is always the

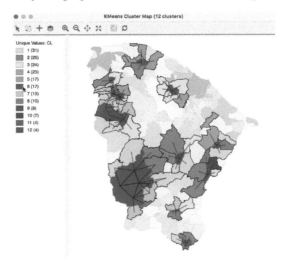

Figure 9.6: Connectivity Check for Cluster Members

midpoint between the current value and the closest previous value to the left or to the right. This process is continued until the change in the weight is less than a given tolerance.

In an ideal situation (highly unlikely in practice), $w_1 = 0$ and the original cluster solution satisfies the spatial constraints. In the worst case, the search yields a weight w_1 close to 1.0, which implies that the attributes are essentially ignored. More precisely, this means that the contiguity constraint cannot be met jointly with the other attributes. Only a solution that gives all or most of the weight to the X-Y coordinates meets the constraint.

In practice, any final solution with a weight larger than 0.5 should be viewed with skepticism, since it diminishes the importance of attribute similarity. Also, it is possible that the contiguity constraint cannot be satisfied unless (almost) all weight is given to the coordinates. This implies that a spatial contiguous solution is incompatible with the underlying attribute similarity for the given combination of variables and number of clusters (k). The spatially constrained clustering approaches covered in the next two chapters address this tension directly.

9.4.1.1 Connectivity check

In order to check whether the elements of a cluster are contiguous, an efficient search algorithm is implemented, essentially a breadth-first search. The algorithm is of complexity O(n), which ensures that it scales well to large data sets.

The search starts at a random node and identifies all of the neighbors of that node that belong to the same cluster. Technically, the node IDs are pushed onto a *stack*. In this process, neighbors that are not part of the same cluster are ignored (i.e., they are not pushed onto the stack). Each element on the stack is examined in turn. If it has neighbors in the cluster that are not yet on the stack, they are added. Otherwise, nothing needs to be changed. After this evaluation, the node is popped from the stack. When the stack is empty, this means that all contiguous nodes have been examined.

At that point, if there are still unexamined observations in the cluster, i.e., if there are cluster members that are not in the current neighbor list, they must be unconnected. Therefore, the cluster is *not* contiguous and the process stops. In general, the examination of nodes

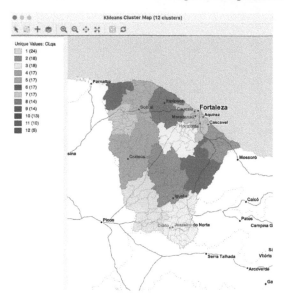

Figure 9.7: Cluster Map, K-Means with Weighted Optimization

is carried out until contiguity fails. If all nodes are examined without failure, all clusters must consist of contiguous nodes (according to the definition used in the spatial weights specification). Since the algorithm is linear in the number of observations, it scales well.

To illustrate this process, consider the partial connectivity graph in Figure 9.6, based on the results for K-Means, shown in the left-hand panel of Figure 9.3. The neighbor structure is shown for the 17 members of C6, based on queen contiguity. This cluster consists of 12 unconnected subclusters, ranging in size from one to five elements.

If the random selection results in one of the singletons being picked, the check would be completed immediately, since those observations only have connections to locations outside the cluster. Since the remaining observations have not been visited, the cluster is clearly not contiguous.

The process is only slightly more complex if the random start is an observation from one of the other groups that consists of multiple elements. In each instance, once the connections to the within group members have been checked, it is clear that the set of 17 observations has not been exhausted, so that the contiguity constraint fails. Once the contiguity constraint is not met for one of the clusters, there is no use in examining other clusters and the search stops.

9.4.2 Implementation

The weighted clustering is invoked in the same manner for each of the five standard clustering techniques. Just below the list of variables in the **Clustering Settings** interface, there is a check box next to **Use geometric centroids**. The default setting is **Auto Weighting**, which will run the bisection search. Alternatively, a **Weighting** can be specified using a slider bar or by entering a numerical value (the default is **1.00**). The coordinates of the spatial units do *not* need to be specified in the variable list. They are calculated under the hood. As before, a warning is generated when the projection information is invalid.

```
Method: KMeans
Number of clusters: 12
Initialization method: KMeans++
Initialization re-runs: 150
Maximum iterations: 1000
Transformation: Standardize (Z)
Distance function: Euclidean
Use geometric centroids (weighting):
    Centroid (X) 0.445008
    Centroid (Y) 0.445008
    mobility 0.0183308
    environ 0.0183308
    housing 0.0183308
    sanitation 0.0183308
    infra 0.0183308
    gdpcap 0.0183308
```

Cluster centers:

	mobility	environ	housing	sanitation	infra	gdpcap
C1	0.963333	0.835	0.842125	0.665292	0.538917	4.58179
C2	0.968722	0.907111	0.813778	0.6565	0.543	4.6885
C3	0.959278	0.847722	0.806333	0.636889	0.464556	4.54378
C4	0.904824	0.791706	0.794294	0.610118	0.468706	10.5602
C5	0.970118	0.870941	0.850882	0.631471	0.538176	4.78641
C6	0.946882	0.874118	0.798294	0.615412	0.529412	4.90559
C7	0.961235	0.833529	0.833647	0.639235	0.551765	3.98812
C8	0.951214	0.889071	0.852643	0.618429	0.487357	4.2305
C9	0.964357	0.887	0.848071	0.717571	0.519429	4.69707
C10	0.964615	0.836615	0.843615	0.694	0.566	5.08031
C11	0.9441	0.9113	0.8513	0.5843	0.414	10.1518
C12	0.9552	0.7336	0.7668	0.5638	0.4994	4.2734

```
The total sum of squares:  1098
Within-cluster sum of squares:
```

	Within cluster S.S.
C1	98.3746
C2	79.4259
C3	59.6847
C4	170.388
C5	41.9099
C6	49.1258
C7	77.0546
C8	39.7148
C9	24.1981
C10	83.4612
C11	87.8312
C12	15.1953

```
The total within-cluster sum of squares:   826.364
The between-cluster sum of squares:    271.636
The ratio of between to total sum of squares:  0.247392
```

Figure 9.8: Cluster Characteristics, K-Means with Weighted Optimization

In addition, a spatial weights file needs to be specified. This is used to assess the spatial constraint. In the absence of a weights file, a warning is generated. In the illustration, the weights file is for queen contiguity. The other six variables are the same as in the previous examples.

To illustrate the process, consider a K-Means clustering application. Before computing the clusters, the **Auto Weighting** button must be clicked to find the solution of the bisection search. In this application, the resulting weight is **0.890015**, clearly much larger than 0.5. As a consequence, the clusters may be spatially contiguous, but they will largely ignore the attribute similarity.

The resulting cluster map is shown in Figure 9.7. The clusters are clearly contiguous and their spatial layout shows a strong resemblance to the coordinate-based clusters in the left-hand panel of Figure 9.1. On the other hand, they do not match the spatial pattern of the pure K-Means result in the left-hand panel of Figure 9.3, except to some extent for the central region and the region around the city of Fortaleza.

The cluster characteristics are given in Figure 9.8. They include the weights for the variables, i.e., 0.445 for each of the coordinates and 0.01833 for each of the six socio-economic attributes. The overall fit of the cluster on the attributes (the coordinates are not included in this calculation) is a pretty dismal BSS/TSS ratio of 0.247, compared to the unrestricted K-Means result of 0.682. This highlights the difficulty in fusing the objectives of attribute similarity and locational similarity.

```
Method: KMeans
Number of clusters: 12
Initialization method: KMeans++
Initialization re-runs: 150
Maximum iterations: 1000
Transformation: Standardize (Z)
Distance function: Euclidean
Use geometric centroids (weighting):
   Centroid (X) 0.125
   Centroid (Y) 0.125
   mobility 0.125
   environ 0.125
   housing 0.125
   sanitation 0.125
   infra 0.125
   gdpcap 0.125

Cluster centers:
    |mobility|environ |housing |sanitation|infra    |gdpcap
----|--------|--------|--------|----------|---------|-------
C1  |0.962962|0.885462|0.833846|0.690692  |0.569692 |4.55896
C2  |0.946913|0.860261|0.785087|0.536     |0.478696 |4.79848
C3  |0.962043|0.892565|0.84987 |0.59      |0.481304 |4.42665
C4  |0.966524|0.909286|0.847714|0.682619  |0.594476 |4.17338
C5  |0.954   |0.70385 |0.84435 |0.6361    |0.5566   |4.9721
C6  |0.966   |0.903947|0.843158|0.804632  |0.548947 |5.36432
C7  |0.951563|0.906125|0.845125|0.593813  |0.4165   |6.18844
C8  |0.975071|0.902929|0.823786|0.609286  |0.445357 |4.164
C9  |0.90975 |0.72825 |0.753875|0.647125  |0.47475  |4.79375
C10 |0.956833|0.715833|0.768   |0.539667  |0.500333 |4.20283
C11 |0.833   |0.76575 |0.818   |0.7925    |0.57375  |12.6005
C12 |0.9525  |0.85925 |0.81225 |0.56375   |0.43125  |27.7873

The total sum of squares:  1098
Within-cluster sum of squares:
    |Within cluster S.S.
----|-------------------
C1  |71.749
C2  |44.9261
C3  |46.7822
C4  |29.6901
C5  |58.7407
C6  |39.2593
C7  |40.5663
C8  |43.3042
C9  |34.3652
C10 |13.1319
C11 |11.0881
C12 |22.227

The total within-cluster sum of squares:   455.83
The between-cluster sum of squares:    642.17
The ratio of between to total sum of squares:  0.584854
```

Figure 9.9: Cluster Characteristics, K-Means with Weights at 0.25

As mentioned in the discussion of the cluster characteristics with an augmented feature set, the X and Y coordinates are included in the computation of the cluster centers and the various measures of fit. The resulting BSS/TSS ratio is therefore not directly comparable with the results from the original K-Means. Since the augmented approach boils down to using equal weights for all variables, the corresponding weights for the coordinates together are $1/8 + 1/8 = 0.250$.

The results of the cluster map in the right-hand panel in Figure 9.3 can be replicated by setting the **Weighting** to **0.25**. The cluster map is identical to the one in Figure 9.3, but the cluster characteristics are not. They are listed in Figure 9.9. The variable weights are 0.125 (1/8) each, and the resulting cluster centers no longer include the information on the X-Y coordinates. The BSS/TSS ratio is now directly comparable to the standard K-Means results. At 0.585, a decrease from 0.682, it shows the price to pay for the spatial nudging, even though the end result is not contiguous.

9.5   Constructing a Spatially Contiguous Solution

A final heuristic to illustrate the tension between attribute similarity and locational similarity constructs a contiguous spatial layout from an initial noncontiguous cluster solution. The heuristic only considers spatial aspects of the data, i.e., contiguity, and ignores the attribute values. Therefore, it will seldom yield an optimal solution. Nevertheless, it serves to illustrate

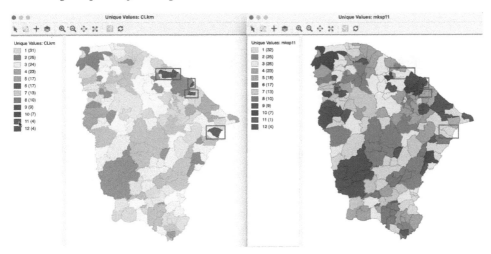

Figure 9.10: Make Spatial Cluster 11

the trade-offs between the two objectives and provides a benchmark to compare to the spatially constrained solutions.

The principle behind the heuristic is to combine unconnected cluster subsets with the largest adjoining subcluster, until all clusters either are contiguous or consist of a singleton, respecting the original number of clusters, k. As usual, the contiguity structure is defined by a spatial weights matrix.

The process starts with the smallest nonsingleton cluster – singletons are left alone. If it consists entirely of contiguous elements, then it is left as is, and the next smallest cluster is considered. A noncontiguous cluster is a cluster that consists of two or more subcomponents. It is evaluated from small to large. Any singleton sub-clusters are assigned to the largest adjoining cluster. The same is done with any larger subcluster, moving from small to large until there is only one entity for that cluster class. When determining which is the largest adjoining cluster, the cluster elements are updated at each step.

The result for each unconnected cluster is that all its smallest elements are absorbed by other subclusters, until only a single entity remains.

To illustrate this process, consider the cluster map for the K-Means solution depicted in the left-hand panel of Figure 9.10. The smallest cluster, C12, consists of a set of four contiguous municipalities, so it does not need to be considered. The next smallest cluster, C11, consists of four singletons, shown as brown cells in the selection in the figure, highlighted within a red rectangle. Each singleton is considered in turn (in random order) and assigned to its largest adjoining cluster, until there is only one left (using queen contiguity to define adjoining cells). In the right-hand panel of the figure, the top-most singleton is assigned to cluster C3 (light green), the middle one to C5 (light rose) and the right-most one to C1 (light blue). The one by the coast is left as is. In this unique values map, the cluster category counts are adjusted accordingly. C11 now has only 1 member, C1 has 32, C3 25 and C5 17 members.

The heuristic then moves to the next smallest cluster, C10, which consists of four disconnected parts: two singletons, one group of two municipalities, and one group of three, as shown in the red rectangles in the left-hand panel of Figure 9.11. The singletons are assigned to the largest adjoining units, respectively C3 for the top-most one, and C2 for the other one. The

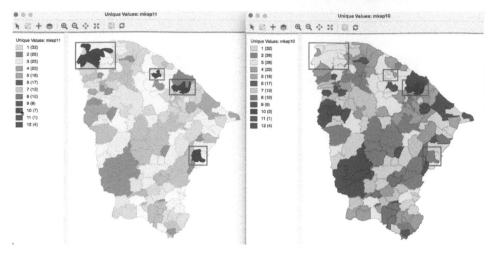

Figure 9.11: Make Spatial Cluster 10

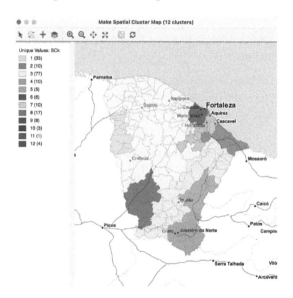

Figure 9.12: Make Spatial Cluster Map

two entities in the north are assigned to C3. The remaining cluster of three units becomes the final cluster 10, as depicted in the right-hand panel.

The process continues until all unconnected entities are assigned.

9.5.1   Implementation

The heuristic can be started from the cluster toolbar icon as the item **Make Spatial** (in the very bottom part of the list) or alternatively as **SC K Means**. The latter is a bit of a misnomer, since it is not actually a spatially constrained K-Means algorithm, but rather the spatial post-processing of a K-Means solution. The **SC K Means** interface is the same as for regular K-Means, except that **Spatial Weights** need to be selected. In this case,

```
Cluster Indicator: SCk
Number of clusters: 12
Weights:     ceara_merc_q
Transformation: Standardize (Z)

Cluster centers:
     |mobility|environ |housing |sanitation|infra   |gdpcap
 --- |--------|--------|--------|----------|--------|-------
 C1  |0.964212|0.845485|0.844636|0.640303  |0.515364|5.741
 C2  |0.959   |0.8571  |0.8254  |0.7622    |0.5882  |4.7095
 C3  |0.961636|0.874857|0.81839 |0.646065  |0.520688|4.95175
 C4  |0.9455  |0.7808  |0.7993  |0.5814    |0.4819  |4.0775
 C5  |0.9504  |0.852   |0.8008  |0.487     |0.4442  |9.4702
 C6  |0.964667|0.908   |0.8585  |0.687     |0.595167|4.03
 C7  |0.9492  |0.8829  |0.8564  |0.6018    |0.4879  |4.6727
 C8  |0.965353|0.827353|0.842941|0.657     |0.526706|4.78847
 C9  |0.9335  |0.90175 |0.841375|0.558     |0.40625 |6.24088
 C10 |0.874   |0.710667|0.742   |0.612     |0.420333|5.307
 C11 |0.936   |0.789   |0.794   |0.545     |0.425   |27.625
 C12 |0.833   |0.76575 |0.818   |0.7925    |0.57375 |12.6005

The total sum of squares:   1098
Within-cluster sum of squares:
     |Within cluster S.S.
 --- |-------------------
 C1  |211.559
 C2  |53.3834
 C3  |331.368
 C4  |24.9483
 C5  |11.5751
 C6  |9.90685
 C7  |25.7885
 C8  |79.0082
 C9  |20.2301
 C10 |5.95639
 C11 |0
 C12 |11.0881

The total within-cluster sum of squares:   784.812
The between-cluster sum of squares:   313.188
The ratio of between to total sum of squares: 0.285235
```

Figure 9.13: Make Spatial Cluster Characteristics

the post-processing is part of the overall calculations and does not need to be carried out separately.

The **Make Spatial Settings** dialog has a slightly different look from the standard cluster interface. The main inputs are **Select Cluster Indicator** and the **Cluster Variables**. The cluster indicator is an integer variable that was saved as the output of a previous clustering exercise. The cluster variables are only included to compute the cluster characteristics at the end. They are not part of the actual heuristic. A **Spatial Weights** file needs to be specified as well (here, queen contiguity). In addition, there is the usual **Transformation** option, again to make sure the cluster characteristics are computed correctly.

The resulting cluster map for K-Means is shown in Figure 9.12. In contrast to the convention for unique values maps, the categories are *not* ordered by size, but they are kept in the same order as for the original cluster map. Consequently, the map in Figure 9.12 has the same cluster categories and colors as the map in Figure 9.10. The number of cluster elements in the final solution is given in parentheses next to the cluster category.

The spatial layout of the clusters shows little resemblance to the other (pseudo-)spatial solutions in this chapter because the centroid coordinates are not part of the heuristic. Instead, it is based on contiguity.

Finally, the cluster characteristics are listed in Figure 9.13. Since the attributes are ignored in the spatial operations, the resulting within-cluster sum of squares are much worse. The overall BSS/TSS ratio is 0.285, similar (and slightly better) to the weighted optimization result.

As mentioned, the approaches outlined in this chapter are primarily for pedagogical purposes, to illustrate the difficult trade-offs between attribute and locational similarity. The results are seldom optimal, or even good, but they highlight the tension between the two objectives. In most applications, there is no easy solution that satisfies both goals.

10

Spatially Constrained Clustering – Hierarchical Methods

In this chapter, the focus shifts to methods that impose contiguity as a *hard* constraint in a clustering procedure. Such methods are known under a number of different terms, including *zonation, districting, regionalization, spatially constrained clustering,* the *p-region problem* and the *p-compact regions problem*. They are concerned with dividing an original set of n spatial units into p internally connected regions that maximize within similarity (for recent reviews, see, e.g., Murray and Grubesic, 2002; Duque et al., 2007, 2011; Li et al., 2014).[1]

In the previous chapter, approaches were considered that impose a *soft* spatial constraint, in the form of a trade-off between attribute similarity and spatial similarity. In the methods considered in this and next chapter, the contiguity is a strict constraint, in that clusters can only consist of entities that are geographically connected. As a result, in some cases the resulting attribute similarity can be of poor quality, when dissimilar units are grouped together primarily due to the contiguity constraint.

Three sets of methods are considered, all based on the principle of hierarchical clustering.

The first approach employs *agglomerative* hierarchical clustering, to which a spatial contiguity constraint is introduced. Early descriptions of the idea of *spatially constrained hierarchical clustering* can be found in Murtagh (1985) and Gordon (1996), among others. Next, an algorithm is considered that takes a *divisive* hierarchical approach. The *SKATER* algorithm (Assunção et al., 2006; Teixeira et al., 2015; Aydin et al., 2018), or, *Spatial 'K'luster Analysis by Tree Edge Removal*, obtains regionalization through a graph partitioning approach. Finally, a collection of methods that combine aspects of both agglomerative and divisive clustering is outlined, referred to as *REDCAP* (Guo, 2008; Guo and Wang, 2011). The acronym stands for *REgionalization with Dynamically Constrained Agglomerative clustering and Partitioning*, and refers to a family of six (later extended to eight) hierarchical regionalization methods.

As before, the methods considered here share many of the same options with previously discussed techniques covered in earlier chapters. Common options will not be considered, but the focus will be on aspects that are specific to the spatially constrained methods.

The *Ceará Zika* sample data set is again used to illustrate the methods, with the same variables as in the previous chapter.

10.1 Topics Covered

- Understand how imposing spatial constraints affects hierarchical agglomerative clustering

---

[1] The terminology is again a bit confusing, since in this literature p is often used for the number of clusters, instead of k, standard for classic clustering techniques. In this chapter, both are employed. The context will make clear how many clusters are considered.

DOI: 10.1201/9781032713175-10

- Understand the tree partitioning method underlying the SKATER algorithm
- Identify contiguous clusters by means of the SKATER algorithm
- Understand the connection between SCHC, SKATER and the REDCAP family of methods
- Identify contiguous clusters by means of the REDCAP algorithm

GeoDa Functions

- Clusters > SCHC
- Clusters > skater
  - set minimum bound variable
  - set minimum cluster size
- Clusters > redcap

Toolbar Icons

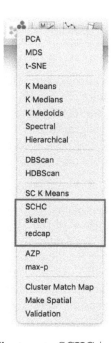

Figure 10.1: Clusters > SCHC | skater | redcap

10.2 Spatially Constrained Hierarchical Clustering (SCHC)

Spatially constrained hierarchical clustering is a special form of constrained clustering, where the constraint is based on contiguity (common borders). In the discussion of classic clustering algorithms, a constrained approach was already considered when a minimum size was imposed. The spatial constraint is more complex in that it directly affects the way in which elemental units can be merged into larger entities.

The idea of including contiguity constraints into agglomerative hierarchical clustering goes back a long way, with early overviews of the principles involved in Lankford (1969), Murtagh (1985) and Gordon (1996). Recent software implementations can be found in Guo (2009) and Recchia (2010), among others.

The clustering logic is identical to that followed for unconstrained hierarchical clustering, and the same expressions are used for linkage and updating formulas, i.e., single linkage, complete linkage, average linkage and Ward's method (see Chapter 5 for details). The only difference is that a contiguity constraint is imposed, based on the nonzero elements of a spatial weights matrix ($w_{ij} \neq 0$).

More specifically, two entities i and j are merged when the dissimilarity measure d_{ij} is the smallest among all pairs of entities, subject to $w_{ij} = 1$. In other words, merger is only carried out when the corresponding entry in the spatial weights (contiguity) matrix is nonzero. Another way to phrase this is that the minimum of the dissimilarity measure is only searched for those pairs of observations that are contiguous.

As before, the dissimilarity measure is recomputed for the newly merged unit using the appropriate formula. In addition, the weights matrix needs to be updated to reflect the contiguity structure of the newly merged units.

In the first step, the weights matrix is of dimension $n \times n$. If observations i and j are merged into a new entity, say A, then the resulting matrix will be of dimension $(n-1) \times (n-1)$ and the two original rows/columns for i and j will be replaced by a new row/column for A. For the new matrix, the row elements $w_{Ah} = 1$ if either $w_{ih} = 1$ *or* $w_{jh} = 1$ (or both are nonzero), and similarly for the column elements.

The next step thus consists of a new dissimilarity matrix and new contiguity matrix, both of dimension $(n-1) \times (n-1)$. At this point, the process repeats itself. As for other hierarchical clustering methods, the end result is a single cluster that contains all the observations. The process of merging consecutive entities is graphically represented in an *dendrogram*, as before.

One potential complication is so-called inversion, when the dissimilarity criterion for the newly merged unit with respect to remaining observations is *better* than for the merged units themselves. This link reversal occurs when $d_{i \cup j, k} < d_{ij}$. It is only avoided in the complete linkage case (for details, see Murtagh, 1985; Gordon, 1996). This problem is primarily one of interpretation and does not preclude the clustering methods from being applied.

10.2.1 The Algorithm

To illustrate the mechanics of the algorithm, a small worked example is provided, using the layout of the 14 counties in the U.S. state of Arizona, shown in Figure 10.2.

To keep matters simple, only a single variable is used, listed in standardized form in Figure 10.3.[2]

A final element is the spatial weights matrix, here implemented as first order queen contiguity, contained in Figure 10.4.

10.2.1.1 SCHC Complete Linkage

SCHC is implemented for four classic linkage functions. To keep the illustration simple, complete linkage is used, which tends to yield compact clusters with an easy updating formula (max distance). Even though this is not an ideal linkage selection, it is easy to implement and helps to illustrate the logic of the SCHC.

---

[2]The variable is the standardized unemployment rate for 1990, from the **natregimes GeoDa** sample data set.

Figure 10.2: Arizona counties

AZID	County	SUE
1	Apache	2.7435
2	Cochise	-0.1365
3	Coconino	-0.3167
4	Gila	-0.0317
5	Graham	0.4561
6	Greenlee	-0.3916
7	La Paz	0.0765
8	Maricopa	-0.9456
9	Mohave	-0.9619
10	Navajo	1.3595
11	Pima	-0.6245
12	Pinal	-0.2877
13	Santa Cruz	-0.0203
14	Yavapai	-0.9192

Figure 10.3: County identifiers and standardized variable

```
0 14 arizona AZID
10 4
3 1 4 5
3 4
10 9 14 4
9 3
3 14 7
1 3
10 6 5
14 5
3 9 4 8 7
4 6
10 3 14 8 5 12
8 5
14 4 12 11 7
6 3
1 5 2
5 7
10 1 4 6 12 11 2
12 4
4 8 5 11
11 6
8 5 12 2 13 7
2 4
6 5 11 13
13 2
11 2
7 4
14 9 8 11
```

Figure 10.4: AZ county queen contiguity (GAL)

	1	2	3	4	5	6	7	8	9	10	11	12	13	14
1	0.00	2.88	3.06	2.78	2.29	3.14	2.67	3.69	3.71	1.38	3.37	3.03	2.76	3.66
2		0.00	0.18	0.10	0.59	0.26	0.21	0.81	0.83	1.50	0.49	0.15	0.12	0.78
3			0.00	0.29	0.77	0.07	0.39	0.63	0.65	1.68	0.31	0.03	0.30	0.60
4				0.00	0.49	0.36	0.11	0.91	0.93	1.39	0.59	0.26	0.01	0.89
5					0.00	0.85	0.38	1.40	1.42	0.90	1.08	0.74	0.48	1.38
6						0.00	0.47	0.55	0.57	1.75	0.23	0.10	0.37	0.53
7							0.00	1.02	1.04	1.28	0.70	0.36	0.10	1.00
8								0.00	0.02	2.31	0.32	0.66	0.93	0.03
9									0.00	2.32	0.34	0.67	0.94	0.04
10										0.00	1.98	1.65	1.38	2.28
11											0.00	0.34	0.60	0.29
12												0.00	0.27	0.63
13													0.00	0.90
14														0.00

Figure 10.5: SCHC complete linkage step 1

The point of departure is the matrix of Euclidean distance in attribute space. In the example, there is only one variable, so the distance is equivalent to the absolute difference between the values at two locations (the square root of the squared difference). In Figure 10.5, the full distance matrix is shown. Distinct from the example in the unconstrained case, attention is limited to those pairs of observations that are spatially contiguous, shown highlighted in red in the matrix.

The first step is to identify the pair of observations that are closest in attribute space and contiguous. Inspection of the distance matrix in Figure 10.5 finds the pair 8-14 as the least different, with a distance value of 0.03.

The next step is to combine observations 8 and 14 and recompute the distance from other observations as the *largest* from either 8 or 14. The updated matrix is shown in Figure 10.6. In addition, the contiguity matrix must be updated with neighbors to the cluster 8-14 as those who were either neighbors to 8 or 14, or to both. The updated neighbor relation is shown as the red values in the matrix. The smallest distance value is between 9 and 8-14, which yields a new cluster of 8-9-14.

	1	2	3	4	5	6	7	9	10	11	12	13	8-14
1	0.00	2.88	3.06	2.78	2.29	3.14	2.67	3.71	1.38	3.37	3.03	2.76	3.69
2		0.00	0.18	0.10	0.59	0.26	0.21	0.83	1.50	0.49	0.15	0.12	0.81
3			0.00	0.29	0.77	0.07	0.39	0.65	1.68	0.31	0.03	0.30	0.63
4				0.00	0.49	0.36	0.11	0.93	1.39	0.59	0.26	0.01	0.91
5					0.00	0.85	0.38	1.42	0.90	1.08	0.74	0.48	1.40
6						0.00	0.47	0.57	1.75	0.23	0.10	0.37	0.55
7							0.00	1.04	1.28	0.70	0.36	0.10	1.04
9								0.00	2.32	0.34	0.67	0.94	0.04
10									0.00	1.98	1.65	1.38	2.31
11										0.00	0.34	0.60	0.32
12											0.00	0.27	0.66
13												0.00	0.93
8-14													0.00

Figure 10.6: SCHC complete linkage step 2

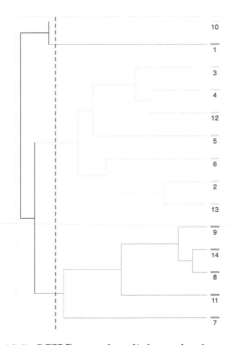

Figure 10.7: SCHC complete linkage dendrogram, k=4

The remaining steps proceed in the same manner. The contiguity relations are updated and the largest distance between the two elements of the cluster is entered as the new distance in the matrix.

The next step combines 2 and 13, followed by 4 and 12, etc.

The complete set of steps can be visualized in a dendrogram, shown in Figure 10.7. As before, the horizontal axis illustrates the change in objective function at each step, starting with the combination of 8 and 14, followed by adding 9, etc.

The spatial layout of the clusters that results for a *cut* at k=4 is shown in Figure 10.8. Clearly, the contiguity constraint has been satisfied.

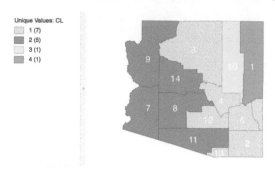

Figure 10.8: SCHC complete linkage cluster map, k=4

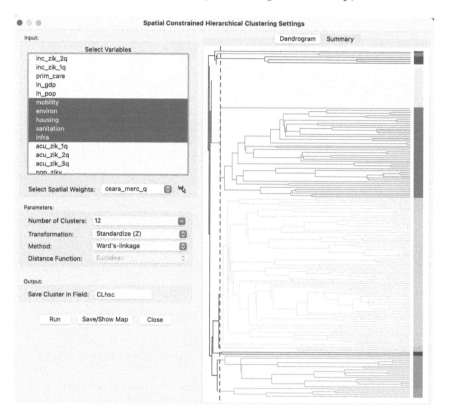

Figure 10.9: SCHC variable selection dialog

10.2.2   Implementation

The example is the same as in the previous chapter, using the 184 municipalities in the Brazilian state of Ceará and six socio-economic indicators: **mobility**, **environ**, **housing**, **sanitation**, **infra** and **gdpcap**. The spatial weights are based on first order queen contiguity.

Only Ward's approach is illustrated, being the preferred linkage method (and the default). The other linkage methods are implemented in the same way, but typically result in worse cluster layouts, similar to what held for their unconstrained counterparts.

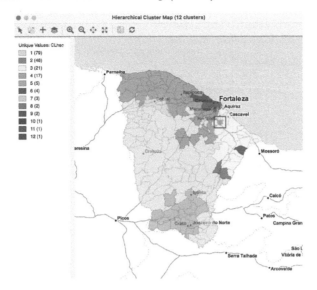

Figure 10.10: SCHC cluster map

Spatially constrained hierarchical clustering is invoked from the drop-down list associated with the cluster toolbar icon, as the first item in the subset highlighted in Figure 10.1. It can also be selected from the menu as **Clusters > SCHC**.

This brings up the **Spatial Constrained Hierarchical Clustering Settings** dialog, which consists of the two customary main panels, as was the case for classic hierarchical clustering.

The settings are shown in Figure 10.9, with the left panel as the interface to specify the variables and a range of parameters. The right panel provides the results, either as a **Dendrogram** or as a tabular **Summary**, each invoked by a button at the top.

In the example, the six variables are selected, with all other items left to their default settings (see Section 5.4.1 for details), except for the location of the cut line, which is set to **12**, consistent with the examples in the previous chapter. Also, the **Method** is kept to **Ward's-linkage**, with the three other linkage methods available from the drop-down list.

The resulting dendrogram is included on the right.

The corresponding cluster map is shown in Figure 10.10. It can be compared to the various cluster maps obtained in the previous chapter. In contrast to those cases, now the clusters all consist of contiguous units. However, the contiguity constraint has resulted in eight clusters out of the 12 containing five or fewer observations, including three singletons. The largest cluster contains almost half the observations (79), the next one is half its size (48) and then again half the size (21). There are also some peculiar configurations that follow from the queen contiguity definition (common vertices). At first sight, the single municipality that belongs to cluster 2 (highlighted in the red square) seems like it should be part of cluster 3. However, it turns out to be connected to two municipalities in cluster 2 with a common vertex.

The loss in cluster performance relative to K-means (see Figure 9.4) is again severe, though not quite as bad as in the examples of the previous chapter. The BSS/TSS ratio is 0.464, relative to 0.682 for K-means (see Figure 10.11).

```
Number of clusters: 12
Transformation: Standardize (Z)
Method: Ward's-linkage
Distance function: Euclidean

Cluster centers:
    |mobility|environ |housing |sanitation|infra   |gdpcap
 ---|--------|--------|--------|----------|--------|-------
 C1 |0.965228|0.891013|0.84643 |0.671595  |0.524481|4.50803
 C2 |0.9515  |0.843646|0.794583|0.5975    |0.503521|4.48277
 C3 |0.950095|0.878524|0.844333|0.575667  |0.440667|6.83433
 C4 |0.951471|0.757529|0.843588|0.649235  |0.588   |4.82082
 C5 |0.9678  |0.8148  |0.7916  |0.5812    |0.4376  |4.2464
 C6 |0.833   |0.76575 |0.818   |0.7925    |0.57375 |12.6005
 C7 |0.992   |0.942333|0.823667|0.869667  |0.625333|6.82333
 C8 |0.8595  |0.687   |0.7215  |0.5835    |0.417   |4.644
 C9 |0.9745  |0.6575  |0.781   |0.8235    |0.693   |5.432
 C10|0.951   |0.824   |0.771   |0.573     |0.443   |25.464
 C11|0.936   |0.789   |0.794   |0.545     |0.425   |27.625
 C12|0.957   |0.966   |0.857   |0.593     |0.357   |40.018

The total sum of squares:  1098
Within-cluster sum of squares:
    |Within cluster S.S.
 ---|--------------------
 C1 |225.732
 C2 |168.248
 C3 |93.8421
 C4 |46.7944
 C5 |18.098
 C6 |11.0881
 C7 |5.59538
 C8 |2.00355
 C9 |16.9901
 C10|0
 C11|0
 C12|0

The total within-cluster sum of squares:  588.391
The between-cluster sum of squares:   509.609
The ratio of between to total sum of squares: 0.464124
```

Figure 10.11: SCHC cluster characteristics

10.3 SKATER

The SKATER algorithm introduced by Assunção et al. (2006), and later extended in Teixeira et al. (2015) and Aydin et al. (2018), is based on the optimal pruning of a minimum spanning tree that reflects the contiguity structure among the observations.[3]

As in SCHC, the point of departure is a dissimilarity matrix that only contains weights for contiguous observations. This matrix is represented as a graph with the observations as nodes and the contiguity relations as edges.

The full graph is reduced to a minimum spanning tree (MST), i.e., such that all nodes are connected (no isolates or islands) and there is exactly one path between any pair of nodes (no loops). As a result, the n nodes are connected by n-1 edges, such that the overall between-node dissimilarity is minimized. This yields a starting sum of squared deviations or SSD as $\sum_i (x_i - \bar{x})^2$, where $\bar{x}$ is the overall mean.

The objective is to reduce the overall SSD by maximizing the between SSD, or, alternatively, minimizing the sum of within SSD. The MST is *pruned* by selecting the edge whose removal increases the objective function (between group dissimilarity) the most. To accomplish this, each potential split is evaluated in terms of its contribution to the objective function.

More precisely, for each tree T, the difference between the overall objective value and the sum of the values for each subtree is evaluated: $SSD_T - (SSD_a + SSD_b)$, where SSD_a, SSD_b are the contributions of each subtree. The contribution is computed by first calculating the

---

[3]This *skater* algorithm is not to be confused with tools to interpret the results of deep learning https://github.com/GapData/skater.

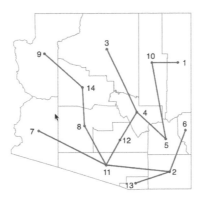

Figure 10.12: SKATER minimum spanning tree

average for that subtree and then obtaining the sum of squared deviations.[4] The cut in the subtree is selected where the difference $\text{SSD}_T - (\text{SSD}_a + \text{SSD}_b)$ is the largest.[5]

At this point, the process is repeated for the new set of subtrees to select an optimal cut. It continues until the desired number of clusters (k) has been reached.

Like SCHC, this is a hierarchical clustering method, but here the approach is divisive instead of agglomerative. In other words, it starts with a single cluster, and finds the optimal split into subclusters until the value of k is satisfied. Because of this hierarchical nature, once the tree is cut at one point, all subsequent cuts are limited to the resulting subtrees. In other words, once an observation ends up in a pruned branch of the tree, it cannot switch back to a previous branch. This is sometimes viewed as a limitation of the SKATER algorithm.

In addition, the contiguity constraint is based on the original configurations and does not take into account new neighbor relations that follow from the combination of different observations into clusters, as was the case for SCHC.

10.3.1 Pruning the Minimum Spanning Tree

The SKATER algorithm is again illustrated by means of the Arizona county example. The first step in the process is to reduce the information for contiguous pairs in the distance matrix of Figure 10.5 (highlighted in red) to a minimum spanning tree (MST). The result is shown in Figure 10.12, against the backdrop of the Arizona county map.

At this point, every possible cut in the MST needs to be evaluated in terms of its contribution to reducing the overall sum of squared deviations (SSD). Since the unemployment rate variable is standardized, its mean is zero by construction. As a result, the total sum of squared deviations is the sum of squares. In the example, this sum is 13.[6]

The vertices for each possible cut are shown as the first two columns in Figure 10.13, with the node number given for the start and endpoint of the edge in the graph. For each subtree,

---

[4]This can readily be computed as $\sum_i x_i^2 - n_T \bar{x}_T^2$, where $\bar{x}_T$ is the average for the subtree, and n_T is the number of nodes in the subtree.

[5]While this exhaustive evaluation of all potential cuts is inherently slow, it can be sped up by exploiting certain heuristics, as described in Assunção et al. (2006). In GeoDa, full enumeration is used, but implemented with parallelization to obtain better performance.

[6]Since the variable is standardized, the estimated variance $\hat{\sigma}^2 = \sum_i x_i^2/(n-1) = 1$. Therefore, $\sum_i x_i^2 = n - 1$, or 13.

MST		SSD	SSD_a	SSD_b	max(ST)
1	10	13.000	0.000	4.894	8.106
2	6	13.000	5.308	0.000	7.692
2	11	13.000	0.072	12.800	0.128
2	13	13.000	12.999	0.000	0.000
3	4	13.000	0.000	12.892	0.108
4	5	13.000	1.527	2.655	8.818
4	12	13.000	6.138	1.345	5.516
5	10	13.000	2.222	0.958	9.820
7	11	13.000	0.000	12.993	0.006
8	11	13.000	0.001	9.609	3.390
8	14	13.000	10.935	0.001	2.064
9	14	13.000	0.000	12.003	0.996
11	12	13.000	1.309	7.202	4.489

Figure 10.13: SKATER step 1

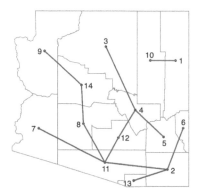

Figure 10.14: SKATER minimum spanning tree – first split

the corresponding SSD must be calculated. The next step then implements the optimal cut such that the total SSD decreases the most, i.e., $\max[SSD_T - (SSD_a + SSD_b)]$, where SSD_T is the SSD for the corresponding tree, and SSD_a and SSD_b are the totals for the subtrees corresponding to the cut.

In order to accomplish this, the SSD for each subtree that would result from the cut is computed as $\sum_i x_i^2 - n_k \bar{x}_k^2$, with $\bar{x}_k$ as the mean for the average value for the subtree and n_k as the number of elements in the subtree.

For example, for the cut 1-10, the vertex 1 becomes a singleton, so its SSD is 0 (SSD_a in the fourth column and first row in Figure 10.13). For vertex 10, the subtree consists of all elements other than 1, with a mean of -0.211 and an SSD of 4.894 (SSD_b in the fifth column and first row in Figure 10.13). Consequently, the contribution to reducing the overall SSD amounts to 13.000 - (0 + 4.894) = 8.106.

In a similar fashion, the SSD for each possible subtree is calculated. This yields the results in Figure 10.13. The largest decrease in overall SSD is achieved by the cut between 5 and 10, which gives a reduction of 9.820.

The updated MST after cutting the link between 5 and 10 is shown in Figure 10.14, yielding two initial subtrees.

The process is now repeated, looking for the greatest decrease in overall SSD. The data is separated into two subtrees, one for 1-10 and the other for the remaining vertices. In each, the SSD for the subtree follows as 0.958 for 1-10 and 2.222 for the other cluster. For each possible cut, the SSD for the corresponding subtrees must be recalculated. The greatest

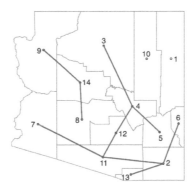

Figure 10.15: SKATER minimum spanning tree – final split

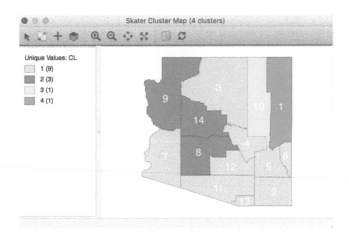

Figure 10.16: SKATER cluster map, k=4

contribution toward reducing the overall SSD is offered by a cut between 8 and 11, with a contribution of 1.441.

At this point, only one more cut is needed (since $k=4$). Neither subtree can match the contribution of 0.958 obtained by 1-10 (since the split yields two single observations, the decrease in SSD equals the total SSD for the subtree). Consequently, no further calculations are needed. The final MST shown in Figure 10.15.

The matching cluster map is given in Figure 10.16. It contains two singletons, a cluster of nine observations and one consisting of three.

10.3.2   Implementation

The SKATER algorithm is invoked from the drop-down list associated with the cluster toolbar icon, as the second item in the subset highlighted in Figure 10.1. It can also be selected from the menu as **Clusters > skater**.

This brings up the **Skater Settings** dialog, which consists of the two customary main panels. As for SCHC, in addition to the variable selection, a spatial weights file must be specified (queen contiguity in the example). The same six variables are used as in the previous examples. The other options are the same as before, with two additional selections: saving

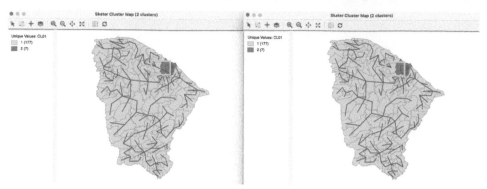

Figure 10.17: Ceará, first cut in MST

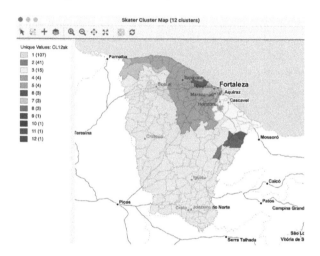

Figure 10.18: Ceará, SKATER cluster map, k=12

the **Spanning Tree** (Section 10.3.2.1), and setting a **Minimum Bound** or **Minimum Region Size** (Section 10.3.2.2).

To illustrate the logic of SKATER, the first cut is highlighted in Figure 10.17, with the MST graph derived from queen contiguity super-imposed on the spatial areal units. The first step in the algorithm cuts the MST into two parts. This generates two sub-regions: one consisting of 177 observations, the other made up of seven municipalities. The green rectangle identifies the edge in the MST where the first cut is made.

The algorithm proceeds by creating subsets of the existing regions, either from the seven observations or from the 177 others. Each subset forms a self-contained sub-tree of the initial MST, corresponding to a region.

To compare the results with the other techniques, Figure 10.18 shows the resulting clusters for k=12. As was the case for SCHC, the layout is characterized by several singletons, and all but three clusters have fewer than five observations.

The overall characteristics of the SKATER result are slightly worse than for SCHC, as depicted in Figure 10.19. The BSS/TSS ratio of 0.4325 is slightly smaller than the 0.4641 for SCHC (compared to 0.682 for unconstrained K-means).

```
Number of clusters: 12
Weights:     ceara_merc_q
Minimum region size:
Distance function: Euclidean
Transformation: Standardize (Z)
```

Cluster centers:

	mobility	environ	housing	sanitation	infra	gdpcap
C1	0.962252	0.860757	0.840215	0.655944	0.531075	4.47473
C2	0.954	0.862098	0.801902	0.615195	0.499293	4.68688
C3	0.944133	0.890267	0.8288	0.548533	0.415067	7.3146
C4	0.833	0.76575	0.818	0.7925	0.57375	12.6005
C5	0.954	0.69125	0.7535	0.533	0.48	4.42375
C6	0.967	0.877667	0.908333	0.706333	0.566333	6.09833
C7	0.992	0.942333	0.823667	0.869667	0.625333	6.82333
C8	0.874	0.710667	0.742	0.612	0.420333	5.307
C9	0.983	0.644	0.865	0.858	0.859	5.578
C10	0.957	0.966	0.857	0.593	0.357	40.018
C11	0.936	0.789	0.794	0.545	0.425	27.625
C12	0.951	0.824	0.771	0.573	0.443	25.464

```
The total sum of squares:  1098
Within-cluster sum of squares:
```

	Within cluster S.S.
C1	398.019
C2	135.799
C3	48.2972
C4	11.0881
C5	9.20636
C6	9.12713
C7	5.59538
C8	5.95639
C9	0
C10	0
C11	0
C12	0

```
The total within-cluster sum of squares:   623.089
The between-cluster sum of squares:    474.911
The ratio of between to total sum of squares:  0.432524
```

Figure 10.19: Ceará, SKATER cluster characteristics, k=12

10.3.2.1 Saving the Minimum Spanning Tree

Figure 10.17 has the graph structure of the MST visualized on the map. In `GeoDa`,this is implemented as a special application of the **Connectivity** property associated with any map window (see Volume 1), with the weights file that corresponds to the MST selected (active) in the **Weights Manager** interface.

The MST connectivity structure is saved by means of the **Save Spanning Tree** option in the **Skater Settings** dialog. One option is to save the **Complete Spanning Tree** (select the corresponding box). The default is to save the MST that corresponds with the current solution, i.e., with the edges removed that follow from the various cuts up to that point.

In Figure 10.17, the full MST is shown in the map on the left, whereas the map on the right shows the MST with one edge removed.

10.3.2.2 Setting a minimum cluster size

In some applications of spatially constrained clustering, the resulting *regions* must satisfy a minimum size requirement. This was discussed in the context of K-means clustering in Section 6.3.4.4. The same option is available here.

There are again two ways to implement the bound. One is based on a spatially extensive variable, like population size. The default is to take the value that corresponds with the 10 percentile, but this does not always lead to feasible solutions. Unlike what is the case for K-means and other not spatially constrained methods, there can be a conflict between the contiguity constraint and the minimum bounds. This is a specific problem for the SKATER algorithm, where solutions for a higher value of k are always subsets of previous solutions. If the resulting subtree does not meet the minimum bound, there is a conflict between the specified k and the desired minimum bound.

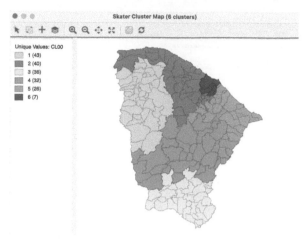

Figure 10.20: Ceará, constrained SKATER cluster map, k=7 with min population size

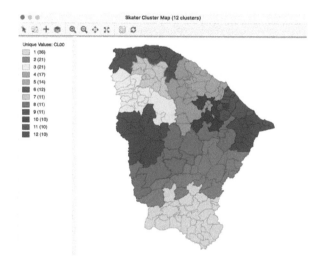

Figure 10.21: Ceará, constrained SKATER cluster map, k=12 with min cluster size 10

The same problem holds for the second minimum bound option, i.e. setting a **Min Region Size**, i.e., the minimum number of observations in each cluster.

For example, with the minimum population set to the 10 percentile of 845,238, it is not possible to obtain seven clusters. As the cluster map in Figure 10.20 shows, the result only shows six clusters, with the smallest consisting of seven observations. The cluster characteristics suffer from the minimum bound constraint, yielding a dismal BSS/TSS ratio of 0.2231.

On the other hand, with a minimum cluster size set to 10, a solution is found for $k=12$, shown in Figure 10.21. The results are much more balanced than the unconstrained solution, with half of the clusters containing 10-12 observations. However, the price paid for this constraint is again a substantial deterioration of the BSS/TSS ratio, to 0.266. Also, this constraint breaks down for values of k of 15 and higher.

10.4   REDCAP

Whereas SCHC involves an agglomerative hierarchical approach and SKATER takes a divisive perspective, the REDCAP collection of methods suggested by Guo (2008) combines the two ideas (see also Guo and Wang, 2011). In this approach, a distinction is made between the linkage update function (originally, *single*, *complete* and *average linkage*), and between the treatment of contiguity.

As Guo (2008) points out, the MST that is at the basis of the SKATER algorithm only takes into account first-order contiguity among pairs of observations. Unlike the approach taken for SCHC, the contiguity relations are not updated to consider the newly formed clusters. As a result of this, observations that are part of a cluster that borders on a given spatial unit are not considered to be neighbors of that unit unless they are also first-order contiguous. The distinction between a fixed contiguity relation and an updated spatial weights matrix is called *FirstOrder* and *FullOrder*. As a result, there are six possible methods, combining the three linkage functions with the two views of contiguity. In later work, Ward's linkage was implemented as well.

In the discussion of density-based clustering in Chapter 20 of Volume 1, it was shown how the single-linkage dendrogram corresponds with a minimum spanning tree (MST). As a result, REDCAP's *FirstOrder-SingleLinkage* and SKATER are identical. Since the *FirstOrder* methods are generally inferior to the *FullOrder* approaches, the latter are the main focus here.[7]

A careful consideration of the various REDCAP algorithms reveals that they essentially consist of three steps. First, a dendrogram for contiguity constrained hierarchical clustering is constructed, using the given linkage function. This yields the exact same dendrogram as produced by SCHC. Next, this dendrogram is turned into a spanning tree, using standard graph manipulation principles. Finally, the optimal cuts in the spanning tree are obtained using the same logic (and computations) as in SKATER, up to the desired level of k.

10.4.1   Illustration – FullOrder-CompleteLinkage

To illustrate the REDCAP *FullOrder-CompleteLinkage* option for the Arizona county data, the results from SCHC (Section 10.2.1.1) and the logic from SKATER (Section 10.3.1) can be reused.

The first stage in the REDCAP algorithm consists of constructing a *spanning tree* that corresponds to a spatially constrained complete linkage hierarchical clustering dendrogram. For *FullOrder*, the steps to follow are the same as for complete linkage SCHC, yielding the dendrogram in Figure 10.7.

To create a spanning tree representation of this dendrogram, observations are connected following the merger of entities in the dendrogram. In the example, this means the first edge is between 8 and 14. In the second step, node 9 is connected to the cluster 8-14, but since 9 is only contiguous to 14, the edge becomes 9-14. The full sequence of edges is given in the list below. Whenever a new entity is connected to an existing cluster, it is connected to the only contiguous unit or to the contiguous unit that is closest (using the distance measures in Figure 10.5):

---

[7]For comparison to SKATER, `GeoDa` includes *FirstOrder-SingleLinkage* as well, but the other *FirstOrder* methods are not implemented.

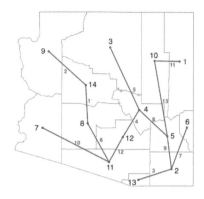

Figure 10.22: REDCAP FullOrder-CompleteLinkage spanning tree

Tree		SSD	SSD_a	SSD_b	max(ST)
1	10	13.000	0.000	4.894	8.106
2	5	13.000	0.072	12.800	0.128
2	6	13.000	5.308	0.000	7.692
2	13	13.000	12.999	0.000	0.000
3	4	13.000	0.000	12.892	0.108
4	5	13.000	1.233	7.075	4.692
4	12	13.000	8.180	0.907	3.912
5	10	13.000	2.222	0.958	9.820
7	11	13.000	0.000	12.993	0.006
8	11	13.000	0.001	9.609	3.390
8	14	13.000	10.935	0.001	2.064
9	14	13.000	0.000	12.003	0.996
11	12	13.000	0.782	8.674	3.543

Figure 10.23: REDCAP step 1

- Step 1: 8-14
- Step 2: 9-14
- Step 3: 2-13
- Step 4: 4-12
- Step 5: 3-4
- Step 6: 11-8
- Step 7: 6-2
- Step 8: 5-4
- Step 9: 2-5
- Step 10: 7-11
- Step 11: 1-10
- Step 12: 11-12
- Step 13: 10-5

The resulting spanning tree is shown in Figure 10.22, with the sequence of steps marked in blue. The tree is largely the same as the MST in Figure 10.12, except that the edge between 11 and 2 is replaced by an edge between 5 and 2.

At this point, an optimal cut in the minimum spanning tree is obtained using the same approach as for SKATER. Given the similarity of the two trees, a lot of the previous results can be reused. Only those subtrees affected by the new edge 5-2 replacing the edge 11-2 need to be considered again. Specifically, in the first step, these are the edges 5-2, 5-4, 4-12 and 11-12. The other results can be borrowed from the first step in SKATER, listed in Figure

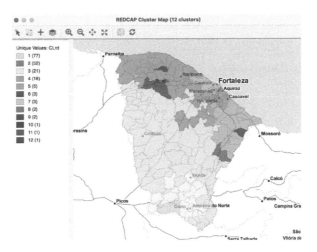

Figure 10.24: Ceará, REDCAP cluster map, k=12

10.13. The new results are given in Figure 10.23. As in the SKATER case, the first cut turns out to be between 5 and 10.

The remaining steps also turn out to be the same as for the SKATER example, yielding the same layout for the four clusters as in Figure 10.16.

10.4.2   Implementation

The REDCAP algorithm is invoked from the drop-down list associated with the cluster toolbar icon, as the last item in the subset highlighted in Figure 10.1. It can also be selected from the menu as **Clusters > redcap.**

Selecting this option brings up the **REDCAP Settings** dialog, which has the same structure as for SKATER, except that multiple methods are available. The panel provides a way to select the variables, the number of clusters and different options to determine the clusters. As for the previous methods, there is a **Weights** drop-down list, where the contiguity weights must be specified. The REDCAP algorithms do not work without a spatial weights file.

Again, as for SKATER, it is possible to impose minimum bounds and to save the minimum spanning tree. These options are not further considered here (see Sections 10.3.2.1 and 10.3.2.2).

With the same six variables as before, the default method of **FullOrder-WardLinkage** (Ward's linkage with dynamically updated spatial weights) yields the cluster map shown in Figure 10.24 for $k=12$.

Compared to the cluster map for SKATER in Figure 10.18, there are some broad similarities, although the layout is different in a number of important respects. The most striking difference is that the largest cluster is quite a bit smaller than before (77 observations vs 107). Also, there are now several sub-clusters in what was the dominant cluster for SKATER. However, there are still three singletons (compared to four) and all but four clusters consist of five or fewer observations.

The cluster characteristics are listed in Figure 10.25. The overall BSS/TSS ratio of 0.4627 is slightly better than SKATER (0.4325), and about the same (but slightly worse) than SCHC (0.4641).

```
Weights:    ceara_merc_q
Method: FullOrder-WardLinkage
Minimum region size:
Save cluster in field: CLrd
Transformation: Standardize (Z)
Distance function: Euclidean
```

Cluster centers:

	mobility	environ	housing	sanitation	infra	gdpcap
C1	0.965753	0.894182	0.84687	0.674688	0.525532	4.52701
C2	0.951269	0.841962	0.792615	0.583019	0.495058	4.97385
C3	0.95219	0.762095	0.839429	0.626333	0.571048	4.64624
C4	0.95	0.886813	0.857938	0.603375	0.439562	6.01063
C5	0.8566	0.7774	0.8086	0.7486	0.5476	15.1732
C6	0.992	0.942333	0.823667	0.869667	0.625333	6.82333
C7	0.969333	0.828667	0.776667	0.631667	0.386667	4.38967
C8	0.8595	0.687	0.7215	0.5835	0.417	4.644
C9	0.9745	0.6575	0.781	0.8235	0.693	5.432
C10	0.958	0.973	0.928	0.798	0.647	3.884
C11	0.936	0.789	0.794	0.545	0.425	27.625
C12	0.957	0.966	0.857	0.593	0.357	40.018

```
The total sum of squares:  1098
Within-cluster sum of squares:
```

	Within cluster S.S.
C1	214.614
C2	173.74
C3	63.0047
C4	66.0625
C5	38.2994
C6	5.59538
C7	9.6022
C8	2.00355
C9	16.9901
C10	0
C11	0
C12	0

```
The total within-cluster sum of squares:  589.912
The between-cluster sum of squares:     508.088
The ratio of between to total sum of squares:  0.46274
```

Figure 10.25: Ceará, REDCAP cluster characteristics, k=12

10.5   Assessment

Clearly, different assumptions and different algorithms yield greatly varying results. This may be discomforting, but it is important to keep in mind that each of these approaches has a slightly different way of handling the tension between attribute and locational similarity.

In the end, the results can be evaluated on a number of different criteria. The ratio of the between sum of squares to the total sum of squares considered so far is only one of a number of possible metrics that can be considered (see Chapter 12). The commonalities and differences between the various approaches highlight where the trade-offs are particularly critical.

Finally, it should be kept in mind that the solutions offered by the different algorithms have no guarantee to yield *global* optima. Therefore, it is important to consider more than one method, in order to gain insight into the sensitivity of the results to the approach chosen.

11

Spatially Constrained Clustering – Partitioning Methods

This third chapter devoted to spatial clustering focuses on the inclusion of explicit spatial constraints in partitioning methods. As mentioned in the previous chapter, in the literature, this is often referred to as the *p-regions problem*.

A discussion of the general issues and extensive literature reviews can be found in Duque et al. (2007), Duque et al. (2011) and Li et al. (2014), among others. Here, the focus of attention is on two specific approaches: AZP and max-p.

In the *automatic zoning problem* or *AZP*, originally considered by Openshaw (1977) (later, AZP is also referred to as the automatic zoning *procedure*), a prior specification of the number of zones or regions (p) is required. In previous chapters, the number of regions was referred to as k, but for consistency with the max-p and p-region terminology, p is used here.

In contrast, in the so-called *max-p regions* model, proposed in Duque et al. (2012), the number of regions becomes endogenous, and heuristics are developed to find the allocation of spatial units into the largest number of regions (max-p), such that a spatially extensive minimum threshold condition is met.

These techniques are again illustrated with the *Ceará Zika* sample data set, using the same six variables and queen contiguity weights as before.

11.1 Topics Covered

- Understand the logic behind the automatic zoning procedure (AZP)
- Appreciate the differences between greedy, simulated annealing and tabu searches
- Gain insight into the different ways to fine-tune AZP using ARiSeL
- Identify contiguous clusters with the number of clusters as endogenous with max-p
- Understand the different stages of the max-p algorithm
- Appreciate the sensitivity of the AZP and max-p heuristics to various tuning parameters

GeoDa Functions

- Clusters > AZP
    - select AZP method
    - ARiSeL option
    - set initial regions
- Clusters > Max-p

DOI: 10.1201/9781032713175-11

Toolbar Icons

Figure 11.1: Clusters > AZP | max-p

11.2 Automatic Zoning Procedure (AZP)

The automatic zoning procedure (AZP) was initially suggested by Openshaw (1977) as a way to address some of the consequences of the modifiable areal unit problem (MAUP). In essence, it consists of a heuristic to find the best set of combinations of contiguous spatial units into p regions, minimizing the within sum of squares as a criterion of homogeneity. The number of regions needs to be specified beforehand, as in all the other clustering methods considered so far.

The problem is NP-hard, so that it is impossible to find an analytical solution. Also, in all but toy problems, a full enumeration of all possible layouts is impractical. In Openshaw and Rao (1995), the original slow hill-climbing heuristic is augmented with a number of other approaches, such as tabu search and simulated annealing, to avoid the problem of becoming trapped in a local solution. None of the heuristics guarantee that a global solution will be found, so sensitivity analysis and some experimentation with different starting points are very important.

Addressing the sensitivity of the solution to starting points is the motivation behind the *automatic regionalization with initial seed location* (ARiSeL) procedure, proposed by Duque and Church in 2004.[1]

---

[1]The origins of this method date back to a presentation at the North American Regional Science Conference in Seattle, WA, November 2004. See the description in the clusterpy documentation at https://github.com/clusterply/clusterpy.

It is important to keep in mind that just running AZP with the default settings is *not sufficient*. Several parameters need to be manipulated to get a good sense of what the best (or, a better) solution might be. This may seem a bit disconcerting at first, but it is intrinsic to the use of a *heuristic* that does not guarantee *global* optimality.

11.2.1 AZP Heuristic

The original AZP heuristic is a local optimization procedure that cycles through a series of possible swaps between spatial units at the boundary of a given region. The process starts with an initial feasible solution, i.e., a grouping of n spatial units into p regions that consist of contiguous units. The initial solution can be constructed in a number of different ways, but it is critical that it satisfies the contiguity constraint.

For example, a solution can be obtained by *growing* a set of contiguous regions from p randomly selected *seed* units by adding neighboring locations until the contiguity constraint can no longer be met. In addition, the order in which neighbors are assigned to growing regions can be based on how *close* they are in attribute space. Alternatively, to save on having to compute the associated WSS, the order can be random. This process yields an initial *list of regions* and an allocation of each spatial unit to one and only one of the regions.

To initiate the search for a local optimum, a random region from the list is selected and its set of neighboring spatial units considered for a swap, one at a time. More specifically, the impact on the objective function is assessed of a move of that unit from its original region to the region under consideration. Such a move is only allowed if it does not break the contiguity constraint in the origin region. If it improves on the overall objective function, i.e., the total within sum of squares, then the move is carried out.

With a new unit added to the region under consideration, its neighbor structure (spatial weights) needs to be updated to include new neighbors from the spatial unit that was moved and that were not part of the original neighbor list.

The evaluation is continued and moves implemented until the (updated) neighbor list is exhausted.

At this point, the process moves to the next randomly picked region from the region list and repeats the evaluation of all the neighbors. When the region list is empty (i.e., all initial regions have been evaluated), the whole operation is repeated with the updated region list until the improvement in the objective function falls below a critical convergence criterion.

The heuristic is local in that it does not try to find the globally best move. It considers only one neighbor of one region at a time, without checking on the potential swaps for the other neighbors or regions. As a result, the process can easily get trapped in a *local* optimum.

11.2.1.1 Illustration

The logic behind the AZP heuristic is illustrated for the Arizona county example, with $p = 4$.

The point of departure is a *random* initial feasible solution with four clusters, for example as depicted in Figure 11.2. The clusters are labeled a (7-9-14), b (1-3-10), c (4-8-11-12) and d (2-5-6-13). Since each cluster consists of contiguous units, it is a *feasible* solution.

The associated within sum of squares for each cluster is computed in Figure 11.3. The values in the column **SUE** are used to compute the cluster average ($\bar{x}_p$), the squared values are listed in column **SUE^2**. For each cluster, the within SSD is calculated as $\sum_i x_i^2 - n_p \bar{x}_p^2$.

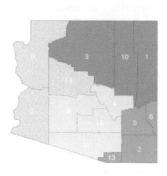

Figure 11.2: Arizona AZP initial feasible solution

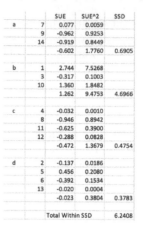

		SUE	SUE^2	SSD
a	7	0.077	0.0059	
	9	-0.962	0.9253	
	14	-0.919	0.8449	
		-0.602	1.7760	0.6905
b	1	2.744	7.5268	
	3	-0.317	0.1003	
	10	1.360	1.8482	
		1.262	9.4753	4.6966
c	4	-0.032	0.0010	
	8	-0.946	0.8942	
	11	-0.625	0.3900	
	12	-0.288	0.0828	
		-0.472	1.3679	0.4754
d	2	-0.137	0.0186	
	5	0.456	0.2080	
	6	-0.392	0.1534	
	13	-0.020	0.0004	
		-0.023	0.3804	0.3783
		Total Within SSD		6.2408

Figure 11.3: Arizona AZP initial Total Within SSD

Figure 11.4: Initial neighbor list

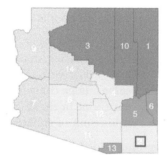

Figure 11.5: Step 1 neighbor selection (2)

For example, for cluster a, this yields $1.7760 - 3 \times (-0.602)^2 = 0.6905$ (rounded). The **Total Within SSD** corresponding to this allocation is 6.2408, listed at the bottom of the figure.

Following the AZP logic, a list of zones is constructed as Z = [a, b, c, d]. In addition, each observation is allocated to a zone, contained in a list, as W = [b, d, b, c, d, d, a, c, a, b, c, c, d, a].

Next, one of the zones is picked *randomly*, e.g., zone c, shown with its labels removed in Figure 11.4. It is then removed from the list, with is updated as Z = [a, b, d]. After evaluating the neighbor list for zone c, this process is repeated for another element of the list, until the list is empty.

Associated with cluster c is a list of its neighbors. These are identified in Figure 11.4 as C = [2, 3, 5, 7, 10, 13, 14].

The next step consists of *randomly* selecting one of the elements of the neighbor list, e.g., 2, and to evaluate its move from its current cluster (b) to cluster c, highlighted by changing its color in the map. However, as shown in Figure 11.5, in this case, swapping observation 2 between b and c would *break* the contiguity in cluster b (13 would become an isolate), so this move is *not allowed*. As a result, 2 stays in cluster b for now.

Figure 11.6: Arizona AZP step 2 neighbor selection – 14

		SUE	SUE^2	SSD
a	7	0.077	0.0059	
	9	-0.962	0.9253	
		-0.443	0.9311	0.5391
b	1	2.744	7.5268	
	3	-0.317	0.1003	
	10	1.360	1.8482	
		1.262	9.4753	4.6966
c	4	-0.032	0.0010	
	8	-0.946	0.8942	
	11	-0.625	0.3900	
	12	-0.288	0.0828	
	14	-0.919	0.8449	
		-0.562	2.2129	0.6351
d	2	-0.137	0.0186	
	5	0.456	0.2080	
	6	-0.392	0.1534	
	13	-0.020	0.0004	
		-0.023	0.3804	0.3783
		Total Within SSD		6.2492

Figure 11.7: Arizona AZP step 2 Total Within SSD

Since the previous move was not accepted, a new move is attempted by randomly selecting another element from the remaining neighbor list C = [3, 4, 5, 7, 10, 13, 14]. For example, observation 14 could be considered for a move from cluster *a* to cluster *c*, as shown in Figure 11.6. This move does not break the contiguity between the remaining elements in cluster *a* (7 remains a neighbor of 9), so it is potentially allowed.

This potential swap would result in cluster *a* consisting of 2 elements and cluster *c* of 5. The corresponding updated sums of squared deviations are given in Figure 11.7. The Total Within SSD becomes 6.2492, which is *not* an improvement over the current objective function (6.2408). As a consequence, this swap is *rejected* and observation 14 stays in cluster *a*.

At this point, observation 14 has been removed from the neighbor list, which becomes C = [3, 4, 5, 7, 10, 13]. A new neighbor is randomly selected, e.g., observation 3. The swap involves a move from cluster *b* to cluster *c*, as in Figure 11.8. This move does not break the contiguity of the remaining elements in cluster *b* (10 and 1 remain neighbors), so it is potentially allowed.

Figure 11.8: Arizona AZP step 3 neighbor selection – 3

		SUE	SUE^2	SSD
a	7	0.077	0.0059	
	9	-0.962	0.9253	
.	14	-0.919	0.8449	
		-0.602	1.7760	0.6905
b	1	2.744	7.5268	
	10	1.360	1.8482	
		2.052	9.3750	0.9577
c	4	-0.032	0.0010	
	8	-0.946	0.8942	
	11	-0.625	0.3900	
	12	-0.288	0.0828	
	3	-0.317	0.1003	
		-0.441	1.4682	0.4948
d	2	-0.137	0.0186	
	5	0.456	0.2080	
	6	-0.392	0.1534	
	13	-0.020	0.0004	
		-0.023	0.3804	0.3783
		Total Within SSD		2.5213

Figure 11.9: Arizona AZP step 3 Total Within SSD

Cluster *b* now consists of 2 elements and cluster *c* has 5. The associated SSD and **Total Within SSD** are listed in Figure 11.9. The swap of 3 from *b* to *c* yields an improvement in the overall objective from 6.2408 to 2.5213, so it is implemented.

The next step is to re-evaluate the list of neighbors of the updated cluster *c*. It turns out that observation 9 must be included as an additional neighbor to the list. The neighbor list thus becomes C = [4, 5, 7, 10, 13, 9].

The process continues by evaluating potential neighbor swaps until the list C is empty. At that point, the original list Z = [a, b, d] is reconsidered and another unit is randomly selected. Its neighbor set is identified and the procedure is repeated until list Z is empty.

After the first full iteration, the whole process is repeated, starting with an updated list Z = [a, b, c, d]. This is carried out until convergence, i.e., until the improvement in the overall objective becomes less than a pre-specified threshold.

11.2.2 Tabu Search

The major idea behind methods to avoid being trapped in a local minimum amounts is to allow nonimproving moves at one or more stages in the optimization process. This purposeful moving in the *wrong* direction provides a way to escape from potentially inferior local optima.

A tabu search is one such method. It was originally suggested in the context of mixed integer programming by Glover (1977), but has found wide applicability in a range of combinatorial problems, including AZP (originally introduced in this context by Openshaw and Rao, 1995).

One aspect of the local search in AZP is that there may be a lot of cycling, in the sense that spatial units are moved from one region to another and at a later step moved back to the original region. In order to avoid this, a tabu search maintains a so-called tabu list that contains a number of (return) steps that are prohibited.

With a given regional layout, all possible swaps are considered from a list of candidates from the adjoining neighbors. Each of these neighbors that is not in the current tabu list is considered for a possible swap, and the best swap is selected. If the best swap improves the overall objective function (the total within sum of squares), then it is implemented. This is the standard approach. In addition, the reverse move (moving the neighbor back to its original region) is added to the tabu list. In practice, this means that this entity cannot be returned to its original cluster for R iterations, where R is the length of the tabu list or the **Tabu Length** parameter in GeoDa.

If the best swap does not improve the overall objective, then the next available tabu move is considered, a so-called aspirational move. If the latter improves on the overall objective, it is carried out. Again, the reverse move is added to the tabu list. However, if the aspirational move does not improve the objective, then the original best swap is implemented anyway and again its reverse move is also added to the tabu list. In a sense, rather than making no move, a move is made that makes the overall objective (slightly) worse. The number of such nonimproving moves is limited by the **ConvTabu** parameter.

The tabu approach can dramatically improve the quality of the end result of the search. However, a critical parameter is the length of the tabu list, or, equivalently, the number of iterations that a tabu move cannot be considered. The results can be highly sensitive to the selection of this parameter, so that some experimentation is recommended (for examples, see the detailed experiments in Duque et al., 2012).

In all other respects, the tabu search AZP uses the same steps as outlined for the original AZP heuristic.

11.2.3 Simulated Annealing

Another method to avoid local minima is so-called *simulated annealing*. This approach originated in physics, and is also known as the Metropolis algorithm, commonly used in Markov Chain Monte Carlo simulation (Metropolis et al., 1953). The idea is to introduce some randomness into the decision to accept a nonimproving move, but to make such moves less and less likely as the heuristic proceeds.

If a move (i.e., a move of a spatial unit into a new region) does not improve the objective function, it can still be accepted with a probability based on the so-called *Boltzmann equation*. It compares the (negative) exponential of the relative change in the objective function to a 0-1 uniform random number. The exponent is divided by a factor, called the *temperature*, which is decreased (lowered) as the process goes on.

Formally, with $\Delta O/O$ as the relative change in the objective function and r as a draw from a uniform 0-1 random distribution, the condition of acceptance of a nonimproving move v is:

$$r < e^{\frac{-\Delta O/O}{T(v)}},$$

where $T(v)$ is the *temperature* at annealing step v.[2] Typically v is constrained so that only a limited number of such annealing moves are allowed per iteration. In addition, only a limited number of iterations are allowed (in `GeoDa`, this is controlled by the **maxit** parameter).

The starting temperature is typically taken as $T = 1$ and gradually reduced at each annealing step v by means of a **cooling rate** c, such that:

$$T(v) = c.T(v-1).$$

In `GeoDa`, the default cooling rate is set to 0.85, but typically some experimentation may be needed. Historically, Openshaw and Rao (1995) suggested values for the cooling rate for AZP between 0.8 and 0.95.

The effect of the cooling rate is that $T(v)$ becomes smaller, so that the value in the negative exponent term $a = \frac{-\Delta O/O}{T(v)}$ becomes larger. Since the relevant expression is e^{-a}, the larger a is, the smaller will be the negative exponential. The resulting value is compared to a uniform random number r, with mean 0.5. Therefore, smaller and smaller values on the right-hand side of the Boltzmann equation will result in less and less likely acceptance of nonimproving moves.

In AZP, the simulated annealing approach is applied to the evaluation step of the neighboring units, i.e., whether or not the move of a spatial unit from its origin region to the region under consideration will improve the objective.

As for the tabu search, the simulated annealing logic only pertains to selecting a neighbor swap. Otherwise, the heuristic is not affected.

11.2.4 ARiSeL

The ARiSeL approach, which stands for *automatic regionalization with initial seed location*, is an alternative way to select the initial feasible solution. In the original AZP formulation, this initial solution is based on a random choice of p seeds, and the initial feasible regions are grown around these seeds by adding the nearest neighbors. It turns out that the result of AZP is highly sensitive to this starting point.

Duque and Church proposed the ARiSeL alternative, based on seeds obtained from a Kmeans++ procedure (see Section 6.2.3). This yields better starting points for growing a whole collection of initial feasible regions. Then the best such solution is chosen as the basis for a tabu search or other search procedure.

11.2.5 Using the Outcome from Another Cluster Routine as the Initial Feasible Region

Rather than using a heuristic to construct a set of initial feasible regions, the outcome of another spatially constrained clustering algorithm could be used. This is particularly

---

[2]The typical application of simulated annealing is a maximization problem, in which the negative sign in the exponential of the Boltzmann equation is absent. However, since in this case the problem is one of *minimizing* the total within sum or squares, the exponent becomes the negative of the relative change in the objective function.

appropriate for hierarchical methods like SCHC, SKATER and REDCAP, where observations cannot be moved once they are assigned to a (sub-)branch of the minimum spanning tree.

GeoDa allows the allocation that resulted from other methods (like spatially constrained hierarchical methods) to be used as the starting point for AZP. This is an alternative to the random generation of a starting point.

An alternative perspective on this approach is to view it as a way to improve the results of the hierarchical methods, where, as mentioned, observations can become trapped in a branch of the dendrogram. The flexible swapping inherent to AZP allows for the search for potential improvements at the margin. In practice, this particular application of AZP has almost always resulted in a better solution, where observations at the edge of previous regions are swapped to improve the objective function.

11.2.6 Implementation

The AZP algorithm is invoked from the drop-down list associated with the cluster toolbar icon, as the first item in the subset highlighted in Figure 11.1. It can also be selected from the menu as **Clusters > AZP**.

The **AZP Settings** interface takes the familiar form, the same as for the other spatially constrained cluster methods. In addition to the variables and the spatial weights, the number of clusters needs to be specified, and the method of interest selected: **AZP** (the default local search), **AZP-Tabu Search** or **AZP-Simulated Annealing**. As before, there is also an option to set a minimum bound.

Two important choices are the search method and the initialization. The latter includes the option to specify a feasible initial region and side-step the customary random initialization. These options are discussed in turn. The other options are the same as before and are not further considered.

The same six variables are used as in the previous chapters, and the number of regions is set to 12 in order to obtain comparable results.

11.2.7 Search Options

The search options are selected from the **Method** drop-down list. Each has its own set of parameters.

To facilitate comparison, recall that the BSS/TSS ratio for K-Means was 0.682, 0.4641 for SCHC, 0.4325 for SKATER and 0.4627 for REDCAP.

11.2.7.1 Local Search

The default option is the local search, invoked as **AZP**. It does not have any special parameter settings. The initialization options are kept to the default.

The resulting cluster map and characteristics are given in Figure 11.10. The clusters are better balanced in size than for the hierarchical solutions and there is only one singleton. However, there remain six clusters with four or fewer observations. The spatial layout is somewhat different from the results obtained so far, although some broad patterns are similar.

The characteristics on the right indicate the method (AZP) and the **Initial value of objective function** as 751.889. This is the WSS, which is reduced to 689.208 in the **Final**

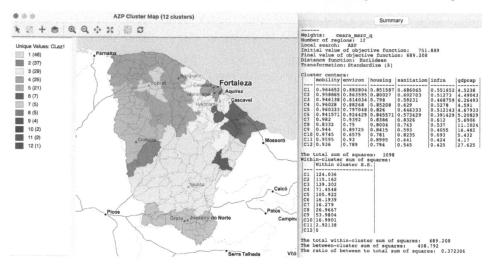

Figure 11.10: Ceará AZP clusters for p=12, local search

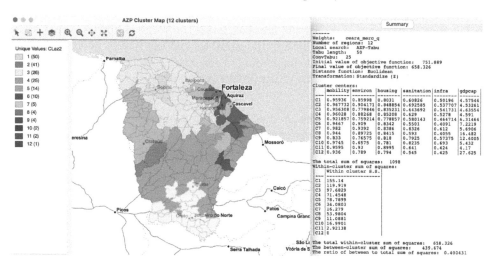

Figure 11.11: Ceará AZP clusters for p=12, tabu search

value. The usual cluster centers are reported, as well as the cluster-specific WSS. The final BSS/TSS ratio is 0.3723, the worst result so far for the spatially constrained methods. As mentioned, the default settings almost never provide a satisfactory result for AZP, and some experimenting is necessary.

11.2.7.2 Tabu search

AZP-Tabu Search is the second item in the **Method** drop-down list. The default parameters are for a **Tabu Length** (10) and for the **ConvTabu**, i.e., the number of nonimproving moves allowed. Initially this is left blank, since it is computed internally, as the maximum of 10 and n/p, which yields 15 in the example. With the default settings, the final value of the objective becomes 689.083 (the initial value is the same as before, since this is not affected by the search method), yielding a BSS/TSS ratio of 0.3724 (not shown).

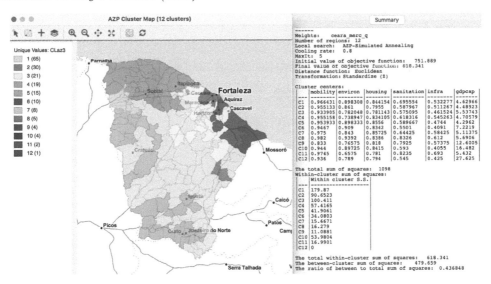

Figure 11.12: Ceará AZP clusters for p=12, simulated annealing search

Some experimenting can improve this to 0.4004 by using a Tabu length of 50 with 25 non-improving moves. This is better than the default AZP, but still worse than the hierarchical results. The cluster map and detailed cluster characteristics are shown in Figure 11.11. The final value of the WSS is 658.326. The spatial layout is again different, with one singleton and six clusters with five or fewer observations. However, closer examination reveals several commonalities with previous results.

11.2.7.3 Simulated annealing

The third option in the **Method** drop-down list is **AZP-Simulated Annealing**. This method is controlled by two important parameters: **Cooling Rate** and **Maxit**. The cooling rate is the rate at which the annealing temperature is allowed to decrease, with a default value of 0.85. **Maxit** sets the number of iterations allowed for each swap, with a default value of 1.

For these default settings, a final BSS/RSS ratio of 0.4146 is obtained, an improvement over the previous results (the final WSS is 642.753). However, the outcome of the simulated annealing approach is very sensitive to the two parameters. Some experimentation reveals a BSS/TSS ratio of 0.4225 for a cooling rate of 0.85 with maxit=5, and a ratio of 0.4368 for a cooling rate of 0.8 with maxit=5. The latter result is depicted in Figure 11.12. The final ratio is now in the same range as the results for the hierarchical methods.

The cluster map contains a singleton (in the higher GDP area focused around the city of Fortaleza), but only two other clusters have four or fewer observations. The overall result is much better balanced than before. However, there are also some strange connections due to common vertices obtained with the queen contiguity criterion. For example, Cluster 1 and Cluster 5 seem to cross to the east of Crateús, and there is a strange connection between Cluster 6 and Cluster 2 south of Horizonte. This illustrates the potential sensitivity of the results to peculiarities of the spatial weights.

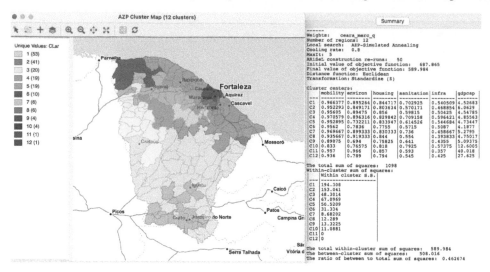

Figure 11.13: Ceará AZP clusters for p=12, simulated annealing search with ARiSeL

11.2.8   Initialization Options

The AZP algorithm is very sensitive to the selection of the initial feasible solution. One easy approach is to assess whether a different random seed might yield a better solution. This is readily accomplished by means of the **Use Specified Seed** check box in the **AZP Settings** dialog.

Alternative is to use the Kmeans++ logic from ARiSeL and to specify the initial regions explicitly.

11.2.8.1   ARiSeL

The **ARiSeL** option is selected by means of a check box in the dialog. There is also the option to change the number of re-runs (the default is 10), which provides additional flexibility. With the default setting and using the best simulated annealing approach (cooling rate of 0.8 with maxit=5), this yields an improved solution with a BSS/TSS ratio of 0.4492. In addition, with the number of re-runs set to 50, the best AZP result so far is obtained (although still not quite as good as SCHC), depicted in Figure 11.13. This yields a BSS/TSS ratio of 0.4627. Note also how the **Initial Value** of the objective function, i.e., the total WSS of the initial feasible solution has improved to 687.865 from the previous 751.889. However, in spite of this better initial solution, the final result is not always better. This very much also depends on the other choices for the algorithm parameters.

The resulting layout shows several similarities with the cluster map in Figure 11.12, although now there are two singletons, and several more small clusters (with less than 10 observations).

11.2.8.2   Initial regions

A final way to potentially improve on the solution is to take a previous spatially constrained cluster assignment as the initial feasible solution. This is accomplished by checking the **Initial Regions** box and selecting a suitable cluster indicator variable from the drop-down list.

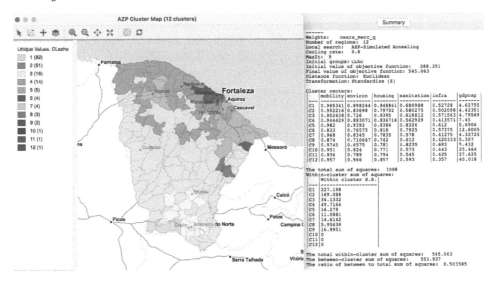

Figure 11.14: Ceará AZP clusters for p=12, simulated annealing search with SCHC initial regions

The approach is illustrated with the SCHC cluster assignment from Section 10.2.2 as the initial feasible solution. It is used in combination with an AZP simulated annealing algorithm with cooling rate 0.8 and maxit=5. The **Initial Value** of the objective function is now 588.391, i.e., the final value from the SCHC solution. This is even slightly better than the final value for the ARiSeL result (589.984), and by far the lowest initial value used up to this point.

The outcome is shown in Figure 11.14. The final BSS/TSS ratio improves to 0.5036, with a final objective value of 545.063, the best result so far. The cluster map has three singletons and is dominated by one large region. The layout is largely the same as for the SCHC solution, with some minor adjustments at the margin.

This illustrates the importance of fine-tuning the various settings for the AZP algorithm, as well as the utility of leveraging several different algorithms.

11.3 Max-P Region Problem

The max-p region problem, outlined in Duque et al. (2012), makes the number of regions (p) part of the solution process. This is accomplished by introducing a minimum size constraint for each cluster. In contrast to the use of such a constraint in earlier methods, where this was optional (see, for example, Section 10.3.2.2), for max-p the *constraint is required*. The size constraint is either a minimum value for a spatially extensive variable (such as population size and number of housing units) or a minimum number of spatial units that need to be contained in each region. Initially formulated as a standard regionalization problem, the method was extended to take into account the compactness of the solutions in Feng et al. (2022), as well as applied to a network structure in She et al. (2017).

In Duque et al. (2012), the solution to the max-p region problem was presented as a special case of mixed integer programming (MIP), building on earlier approaches to incorporate contiguity constraints in site design problems given by Cova and Church (2000). Two key aspects are the formal incorporation of the contiguity constraint and the use of an objective function that combines the number of regions (p) and the overall homogeneity (minimizing the total with sum of squares). However, within this objective function, priority is given to finding a larger p rather than the maximum homogeneity. In contrast to an unconstrained problem, where a larger p will always yield a better measure of homogeneity (see the discussion of the elbow graph for K-means clustering in Section 6.3.4.2), this is not the case when both a minimum size constraint *and* contiguity are imposed. In other words, simple minimization of the total within sum of squares may not necessarily yield the maximum p. Also, the maximum p solution may not give the best total within sum of squares. The trade off between the two sub-objectives is made explicit.

In a nutshell, the max-p algorithm will find the largest p that accommodates both the minimum size and contiguity constraints and then finds the regionalization that yields the smallest total within sum of squares.

As it turns out, the formal MIP strategy becomes impractical for any but small-size problems. Instead, a *heuristic* was proposed that consists of three important steps: growth, assignment of enclaves and spatial search.[3] As was the case for previous heuristics, this does not guarantee that a global optimum is found. Therefore, some experimentation with the various solution parameters is recommended. Even more than in the case of AZP, sensitivity analysis and the tuning of parameters are critical to finding better solutions. Simply running the default will not be sufficient. The trade-offs involved are complex and not always intuitive.

11.3.1 Max-p Heuristic

The max-p heuristic outlined in Duque et al. (2012) consists of three important phases: growth, assignment of enclaves and spatial search.

The objective of the first phase is to find the largest possible value of p that accommodates both the minimum bounds and the contiguity constraints. In order to accomplish this, many different spatial layouts are *grown* by taking a random starting point and adding neighbors (and neighbors of neighbors) until the minimum bound is met. While it is practically impossible to consider all potential layouts, repeating this process for a large number of tries with different starting points will tend to yield a high value for p (if not the highest), given the constraints.

In the process of growing the layout, some spatial units will not be allocated to a region, when they and/or their neighbors do not meet the minimum bounds constraint or break the contiguity requirement. Such spatial units are stored in an *enclave*.

At the end of the growth process, the result consists of one or more spatial layouts with p regions, where p is the largest value that could be obtained. Before proceeding with the optimization process, any remaining spatial units that are contained in the enclave must be assigned to an existing region. This is implemented in such a way that the overall within sum of squares is minimized.

Once all units have been assigned to one of the p regions, the best solution is selected as the *initial feasible solution*. At this point, the heuristic proceeds in exactly the same way as for AZP, using either greedy search, tabu search or simulated annealing to find a (local) optimum. The best overall solution is chosen at the end.

---

[3]Further consideration of some computational aspects is given in Laura et al. (2015) and Wei et al. (2021).

AZID	County	PO90
1	Apache	61,591
2	Cochise	97,624
3	Coconino	96,591
4	Gila	40,216
5	Graham	26,554
6	Greenlee	8,008
7	La Paz	120,739
8	Maricopa	2,122,101
9	Mohave	93,497
10	Navajo	77,658
11	Pima	666,880
12	Pinal	116,379
13	Santa Cruz	29,676
14	Yavapai	107,714

Figure 11.15: Arizona counties population

Figure 11.16: Arizona max-p growth, start with 6

11.3.1.1 Illustration

To continue the illustration of the heuristic by means of the Arizona county example, the population in 1990 (**PO90**) is added as the minimum size constraint. Specifically, a lower limit of a population size of 250,000 is imposed (without any substantive meaning, purely to illustrate the logic of the algorithm). In Figure 11.15, the population count is shown for each county. The other variables are as before (see Section 10.2.1).

The heuristic consists of three phases. In the first, the *growth* phase, feasible regions are created and the largest possible p identified. For those configurations that achieve the maximum p, any unassigned spatial units, the so-called *enclave*, are allocated to existing regions so as to minimize the total within sum of squares. Finally, the best of these solutions is taken as a feasible initial solution to start the *optimization* process. The latter is the same as for AZP.

The *growth* process is initiated by randomly selecting a location and determining its neighbors. In Figure 11.16, county 6 is chosen. Its three neighbors are 1, 5 and 2.

County 6 has a population of 8,008. Its neighbor with the largest population is county 2, with a population of 97,624. Joined with 6 it forms the core of the first region. Its total population is now 105,632, which does not meet the minimum threshold. At this point, the neighbors of 2 that are not already neighbors of 6 are added to the neighbor list.

As shown in Figure 11.17, the neighbor set now also includes 11 and 13.

Figure 11.17: Arizona max-p growth phase – 6 and 2

Figure 11.18: Arizona max-p growth phase – region 1

Figure 11.19: Arizona max-p growth phase – regions 1 and 2

Of the four neighbors, county 11 has the largest population (666,880), which brings the total regional population to 764,504, well above the minimum bound. At this point, the first region becomes 6-2-11, shown in Figure 11.18.

The second random seed yields county 8. Its population is 2,122,101, well above the minimum bound. Therefore, it constitutes the second region in the growth process as a singleton, shown in Figure 11.19.

The third random seed is county 3, with neighbors 9, 14, 4 and 10, shown in Figure 11.20. The population of county 3 is 96,591 and the largest neighbor is 14, with a population of 107,714. This yields a total of 204,305, insufficient to meet the minimum bound.

Figure 11.20: Arizona max-p growth phase – pick 3

Figure 11.21: Arizona max-p growth phase – join 3 and 14

Figure 11.22: Arizona max-p growth phase – region 1, 2, 3

Therefore, again counties 3 and 14 are grouped and the list of neighbors updated, shown in Figure 11.21. The only new neighbor is county 7, since county 8 has already been allocated to the second region.

Since county 7 has the largest population among the eligible neighbors (120,739) it is grouped with counties 3 and 14. This new region meets the minimum threshold, with a total population of 325,044. At this point, there are three regions ($p = 3$), shown in Figure 11.22.

The next random seed yields county 9, with a population of 93,497. Its neighbors are already part of a previously assigned region (see Figure 11.22), and because it has insufficient

Figure 11.23: Arizona max-p growth phase – pick 12

Figure 11.24: Arizona max-p growth phase – join 12 and 4

Figure 11.25: Arizona max-p enclaves and initial regions 1, 2, 3, 4

population, county 9 cannot be turned into a region itself. Therefore it is relegated to the *enclave*.

The following random pick yields county 12, with a population of 116.379, and with as neighbors counties 4, 8, 11 and 5, as shown in Figure 11.23. Of the neighbors, counties 8 and 11 are already part of other regions, so they cannot be considered.

Of the two neighbors of county 12, county 4 has the largest population, at 40,216, so it is joined with county 12. However, their total population of 156,595 does not meet the threshold, so additional neighbors need to be considered as. Of the six neighbors of county

cluster 1	SUE	SUE^2	SSD		cluster 4	SUE	SUE^2	SSD	Total
2	-0.137	0.0186			4	-0.032	0.0010		
6	-0.392	0.1534			10	1.360	1.8482		
11	-0.625	0.3900			12	-0.288	0.0828		
13	-0.020	0.0004				0.347	1.9320		
	-0.293	0.5624	0.2185					1.5714	1.7899
1	2.744	7.5268			5	0.456	0.2080		
	0.314	8.0892	7.5958			0.374	2.1400	1.5804	9.1762
5	0.456	0.2080			1	2.744	7.5268		
	-0.143	0.7704	0.6677			0.946	9.4588	5.8799	6.5476
1	2.744	7.5268							
5	0.456	0.2080							
	0.338	8.2972	7.6126					1.5714	9.1840
					1	2.744	7.5268		
					5	0.456	0.2080		
			0.2185			0.848	9.6668	6.0718	6.2903

Figure 11.26: Arizona max-p assign enclaves

4, that can be included in the neighbor list, only county 10 is an addition, shown in Figure 11.24.

Between counties 5 and 10, the latter has the largest population (120,739), which brings the total for the region 12-4-10 to 277,334. The new configuration with four regions is shown in Figure 11.25, which now sets $p = 4$.

There are four remaining counties. County 9 is already in the enclave. Clearly, county 13 with a population of 29,676 is not large enough to form a region by itself, so it is added to the enclave as well.

Similarly, the combination of counties 5 and 1 only yields a total population of 88,145, which is insufficient to form a region, so they are both included in the enclave.

At this point, the growth phase has been completed and yielded $p = 4$, with four counties left in the enclave, as in Figure 11.25. In a typical application, this process would be repeated many times and only the solutions would be kept that achieve the largest p. For those layouts, the best feasible initial solution must be obtained, in which all units are assigned to regions. Therefore, the counties that remain in the enclave must be assigned to one of the existing regions such as to minimize the total within sum of squares.

In the example, the solution is straightforward for counties 9 and 13: county 9 is merged with 3-14-7, and county 13 is merged with 11-2-6. For counties 5 and 1, the situation is a bit more complicated, since both could be assigned to either cluster 1 (2-6-11-13) or cluster 4 (4-10-12), or one to each, for a total of four combinations.

The calculations of the respective sum of squared deviations are listed in Figure 11.26. The starting point is a total SSD between cluster 1 and 4 of 1.7899, at the top of the figure. Next, the new SSD is computed for the case where county 1 is added to the first cluster and county 5 to the fourth and vice versa. This increases the total SSD between the two clusters to respectively 9.1762 and 6.5476. Finally, the scenarios where both county 1 and county 5 are added either to the first or to the fourth cluster are considered. The results for the new SSD are, respectively, 9.1840 and 6.2903.

Consequently, the allocation that obtains the smallest within sum of squares is the regional configuration that assigns both counties 1 and 5 to region 4, shown in Figure 11.27. The constituting regions are 11-13-2-6, 8, 3-14-7-9 and 1-10-4-12-5.

At this point, there is a feasible initial solution that can form the starting point for one of the three search algorithms, in the same fashion as for AZP.

Figure 11.27: Arizona max-p – feasible initial regions

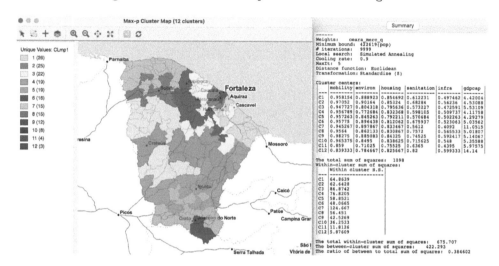

Figure 11.28: Ceará max-p regions with population bound 422,619

11.3.2   Implementation

The max-p option is available from the main menu as **Clusters > max-p**, or from the drop-down list associated with the cluster toolbar icon, as the second item in the subset highlighted in Figure 11.1.

The variable settings dialog is again largely the same as for the other cluster algorithms, with the additional requirement to select a **Mininum Bound** variable and specify a threshold value. The same three **Local Search** options are available as for AZP. The default **#iterations** is set to 99.

From the drop-down list, the variable for the **Minimum Bound** is selected as **pop**, the total population in each municipality. Its default setting is 10% of the total over all observations (i.e., the total population of the state of Ceará), which amounts to 845,238.

With all the default settings (and **Greedy** as the local search), the max-p solution only yields 6 regions, with a dismal BSS/TSS ratio of 0.2756. The main culprit is the highly skewed distribution for population size, with a median population of 19,970 for the municipalities. However, eight municipalities have a population greater than 100,000. The default setting is thus not a reasonable population threshold.

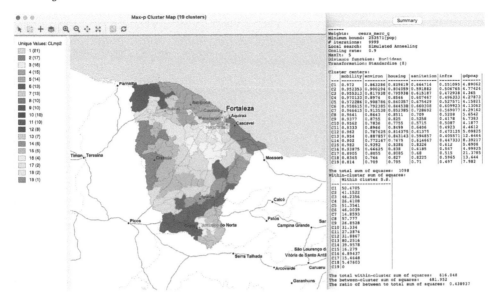

Figure 11.29: Ceará max-p regions with population bound 253,571

Some experimenting suggests a bound of 5% (422,619). With the number of iterations set to 9999 (higher values tend to yield better solutions), and using simulated annealing for the search, with a cooling rate of 0.9 and maxit=5 yields a max-p of 12. This turns out (purely by coincidence) to be the same number of regions as used in the previous examples, which makes the results comparable.

The overall fit is somewhat poorer, with a BSS/TSS ratio of 0.3846. In essence, this is the price to pay for imposing a minimum bound on the solution. The results are shown in Figure 11.28. In contrast to previous layouts, the cluster map is much more balanced, with only two smaller regions and no singletons.

11.3.3 Sensitivity Analysis

The results of max-p are very sensitive to several parameters. Not only does the type of local search affect the results, but the number of iterations is very important, as is the minimum bound. When the latter is not based on a substantive concern (such as legal mandates), experimenting with different bounds will provide insight into the maximum p that can be obtained. For example, with the bound set at 3% of the total population, or 253,571, 19 clusters can be obtained (using simulated annealing with cooling rate 0.9, maxit=5 and 9999 iterations), shown in Figure 11.29. This yields a BSS/TSS ratio of 0.4389.

The solution of spatially constrained partitioning cluster methods is based on heuristics, which require substantial fine-tuning. While this may seem disconcerting at first, it also provides the opportunity to gain closer insight into the various trade-offs involved. These include not only the tension between locational and attribute similarity, but also between the spatial characteristics of the resulting cluster map and the quantitative measures of fit. This is considered more closely in the final chapter.

Part IV

Assessment

12

Cluster Validation

The journey through the range of clustering methods may come across as a bit bewildering, since the choice of algorithm, tuning parameter settings and other decisions lead to sometimes very different results. These so-called *researcher degrees of freedom* (Gelman and Loken, 2014), considered in the Epilogue of Volume 1 are part and parcel of the *unsupervised* aspect of spatial data science.

Unless a given classification is known as *truth*, to which different solutions can be compared, it is near impossible to select a *best* approach. Different criteria will favor some solutions over others. It is therefore important to assess which criteria are most appropriate in a given empirical situation. In this volume, a lot of attention has focused on the tension between attribute similarity and locational similarity. Depending on the context, one will prevail over the other. Even when internal similarity is the dominant objective, it remains important to put the results in a *spatial* context, to assess the spatial distributional aspects of the grouping of observations.

In this final chapter, attention shifts to assessing the performance of a given cluster result to a given standard and how to compare results obtained through different methods to each other. This is referred to in the literature as, respectively, *external validity* and *internal validity* (Akhanli and Hennig, 2020).

External validity is typically used to compare the outcome of various clustering methods to a known *truth*. Unless one is carrying out an experimental design, this is not really that practical in actual empirical situations. However, it remains a very useful approach when studying the trade-offs created by different tuning parameters and other design choices. An extensive review of external validity indices is contained in Meila (2015). In a recent application, Aydin et al. (2021) computed some 13 different metrics to compare the performance of six spatially constrained clustering methods in an experimental setting (true cluster categories known).

In an actual empirical application, measures of internal validity of a cluster are much more useful. As mentioned, such measures tend to favor one objective over another or implement a given compromise between conflicting goals. Typical properties considered are within cluster homogeneity, between cluster separation and stability (for recent overviews of internal validity indices, see Halkidi et al., 2015; Akhanli and Hennig, 2020). Such measures are then often included in a search for the optimal number of clusters or k.

In this chapter, an overview is presented of a range of validity measures, but in the limited context where the value of k is taken as given. This is illustrated for the various results obtained in previous chapters using the *Ceára Zika* sample data set. For consistency, k was always set to 12.

In addition to classic indicators from the literature, some novel *spatial* aspects are introduced as well. This includes a *join count ratio* to assess cluster contiguity structure, and a *cluster match map* to compare the spatial alignment of observations between clusters.

The chapter closes with some brief concluding remarks.

12.1   Topics Covered

- Understand the difference between internal and external validation measures
- Interpret measures of fit
- Assess cluster balance by means of entropy and Simpson's index
- Assess cluster contiguity structure by means of the join count ratio
- Measure the compactness of spatially constrained clusters by means of the isoperimeter quotient
- Assess the cluster connectedness by means of the graph diameter
- Compare different clusterings by means of the Adjusted Rand index (ARI) and the Normalized Information Distance (NID)
- Visualize the spatial overlap between clusters by means of linked cluster maps
- Interpret a cluster match map to compare the spatial pattern in two clusterings

GeoDa Functions

- Clusters > Validation
- Clusters > Cluster Match Map

Toolbar Icons

Figure 12.1: Clusters > Cluster Match Map | Make Spatial | Validation

12.2   Internal Validity

Indices pertaining to the internal validity of a cluster focus on the composition of the clusters in terms of how homogeneous the observations are, how well clusters find the right separation between groups, as well as the balance among individual cluster sizes.

Five types of measures are considered here. First are traditional measures of *fit* that were included among the **Summary** of cluster characteristics in the preceding chapters. Next follow indicators of the *balance* of cluster sizes, i.e., the evenness of the number of observations in each cluster.

The final three types of measures are less common. The *join count ratio* is an indicator of the compactness and separation between clusters in terms of the number of connections with observations outside the cluster. *Compactness* is a characteristic of the shape of a spatially constrained cluster and can be assessed by many different indicators. Finally, *connectedness* is an alternative measure of compactness that is derived from the graph structure of the spatial weights.

12.2.1 Traditional Measures of Fit

As covered in previous chapters, the total sum of squared deviations from the mean (TSS) is decomposed into one part attributed to the within sum of squares (WSS) and a complementary part due to the between sum of squares (BSS). These are the most commonly used indicators of the fit of a clustering to the data. A *better* cluster is considered to be one with a smaller WSS, or, equivalently, a larger BSS to TSS ratio. As noted earlier, this is only an appropriate metric when the dissimilarity matrix is based on an Euclidean distance. When this is not the case, e.g., for K-Medoids, a different metric should be used.

While useful, these indicators of fit miss other important characteristics of a cluster. For example, in order to better identify critical assignments within a given cluster alignment, Kaufman and Rousseeuw (2005) introduced the notion of *average silhouette width*. This is the average dissimilarity of an observation to members of its own cluster compared to the average dissimilarity to observations in the closest cluster to which it was not classified. An extension that takes the spatial configuration into account in the form of so-called geo-silhouettes is presented in Wolf et al. (2021).

12.2.2 Balance

In many applications, it is important that the clusters are of roughly equal size. Two common indicators are the *entropy* and *Simpson's index*. Both compare the distribution of the number of observations among clusters to an even distribution. Whereas entropy is maximized for such a distribution, Simpson's index is minimized.

The entropy for a given clustering P consisting of k clusters with observations n_i in cluster i, for a total of n observations, is (see, e.g., Vinh et al., 2010):

$$H(P) = -\sum_{i=1}^{k} \frac{n_i}{n} \log \frac{n_i}{n}.$$

For equally balanced clusters, $n_i/n = 1/k$, so that entropy is $-k.(1/k)\ln(1/k) = ln(k)$, which is the maximum that can be obtained for a given k. To facilitate comparison among clusters of different sizes, a standardized measure can be computed as entropy/max entropy. The closer this ratio is to 1, the better the cluster balance (by construction, the ratio is always smaller than 1).

The entropy fragmentation index can be computed for the overall cluster result, but also for each cluster separately. This is especially relevant when a cluster consists of subclusters, i.e., subsets of contiguous observations. A cluster that consists of subclusters of equal size would have a relative entropy of 1. The sub-cluster entropy is only a valid measure for noncontiguous or noncompact clusters, so it is not appropriate for the results of spatially constrained clustering methods.

Another indicator of cluster balance is Simpson's index of diversity (Simpson, 1949), also known in economics as the Herfindahl-Hirschman index:

$$S = \sum_{i=1}^{k} (\frac{n_i}{n})^2.$$

With equal representation in clusters, Simpson's index equals $1/k$, the lowest possible value. It ranges from $1/k$ to 1 (all observations in a single cluster). A standardized index yields a value of 1 for the equal representation case, and values larger than 1 for others, with smaller values indicating a more balanced distribution.[1]

Same as for the entropy measure, Simpson's index is only applied to subclusters in cases where the solution is not spatially constrained.

12.2.3   Join Count Ratio

An index that addresses both compactness and separation is the *join count ratio*. It is derived from the contiguity structure among observations that is reflected in a spatial weights matrix. For a given such structure, a relative measure of compactness of a cluster is indicated by how many neighbors of each observation in the cluster are also members of the cluster. For a spatially compact cluster, ignoring border effects, this ratio should be 1. The higher the ratio, the more compact and self-contained the cluster, with the least connectivity to other clusters.

The join count ratio can be computed for each cluster separately, as well as for the clustering as a whole. A value of zero (the minimum) indicates that all neighbors of the cluster members are outside the cluster. This is only possible when the cluster is not spatially constrained and when all cluster elements are singletons in a spatial sense.

12.2.4   Compactness

For spatially constrained clusters, compactness is a key characteristic. For example, this is a legal criterion in the context of electoral redistricting, which is a form of spatially constrained clustering (Saxon, 2020). However, there is no single measure to characterize compactness, and many different aspects of the shape of clusters can be taken into account (e.g., the review in Niemi et al., 1990). For example, Saxon (2020) reviews no less than 18 indicators of compactness and compares their properties in the context of gerrymandering.

Perhaps the most famous measure of compactness is the isoperimeter quotient (IPQ), i.e., the ratio of the area of a cluster shape to that of a circle of equal perimeter (Polsby and Popper, 1990).

The point of departure is the view that a circle is the most compact shape. It is compared to an irregular polygon with the same perimeter.[2] The area of a circle, expressed in function of its perimeter p is $C = p^2/4\pi$.[3] Consequently, the isoperimeter quotient as the ratio of the

---

[1]In the literature, a normalized Herfindahl-Hirschman index is sometimes used as (HHI - 1/n) / (1 - 1/n), which runs from 0 to 1, with 0 as best.

[2]Technically speaking, determining the perimeter of an irregular polygon is not without problems, and depends on the precision of the digital representation of the boundary.

[3]The perimeter of a circle is $p = 2\pi r$, with r as the radius, so $r = p/2\pi$. The area of a circle is $C = \pi r^2$, so expressed as a function of the perimeter, it is $C = \pi(p/2\pi)^2 = p^2/4\pi$.

area of the polygon A over the area of the circle is:

$$IPQ = 4\pi A/p^2,$$

with A as the area of the polygon.

The IPQ is only suitable for spatially constrained cluster results.

12.2.5 Connectedness

With the spatial weights viewed as a network or graph, a spatially constrained cluster must constitute a so-called connected component. The diameter of the network structure that corresponds with the cluster is the length of the longest shortest path between any pair of observations (Newman, 2018, p. 133). Starting from an unweighted graph representation of the spatial weights matrix, each connection between two neighbors corresponds with one step in the graph.

For a given number of observations in a cluster, the diameter computed from the spatial weights connectivity graph gives a measure of compactness (smaller is more compact). For example, for a star-shaped layout of observations, the diameter would equal two (the longest shortest path between any pair of observations goes through the center of the star in two steps). On the other end of the spectrum, for a long string of m observations, the diameter would be $m - 1$.

Everything else being the same, the diameter of a network increases with its size. Dividing the diameter by the number of observations in the cluster gives a relative measure, which corrects for the size of the cluster.

As for the IPQ, the diameter of a cluster is only applicable to spatially constrained clusters.

12.2.6 Implementation

To compare the different cluster validation measures, eight different outcomes are considered, all obtained with $k = 12$ for the Ceará economic indicators ($n = 184$). Two clusterings are nonspatial, i.e., Hierarchical clustering with Ward's linkage (not used earlier, but shown in Figure 12.2) and K-Means (cluster map in the left-hand panel of Figure 9.3, characteristics in Figure 9.4). The remaining six patterns represent different methods to obtain spatially constrained results: SCHC with Ward's linkage (Figure 10.10 and Figure 10.11); SKATER (Figure 10.18 and Figure 10.19); REDCAP (Figure 10.24 and Figure 10.25); AZP with simulated annealing (Figure 11.12); AZP with SCHC as initial solution (Figure 11.14) and the max-p outcome that yielded $p = 12$ (Figure 11.28).

The WSS, BSS/TSS ratio, overall Entropy, Simpson's index and the join count ratio for each clustering are listed in Figure 12.3. The first two measures are included as part of the **Summary** for each cluster result. The others are invoked by selecting **Clusters > Validation** from the menu or as the last item in the cluster drop-down list from the toolbar (see Figure 12.1). The required input is a **Cluster Indicator** (i.e., the categorical variable saved when carrying out a clustering exercise) and a spatial weights file. The latter is required for the join count ratio, even for traditional (nonspatial) clustering methods.

The **Validation** option brings up a results window, shown in Figure 12.4 for the hierarchical clustering outcome. At the top, this gives the number of clusters (12), the raw and standardized entropy measures as well as the raw and standardized Simpson's index.

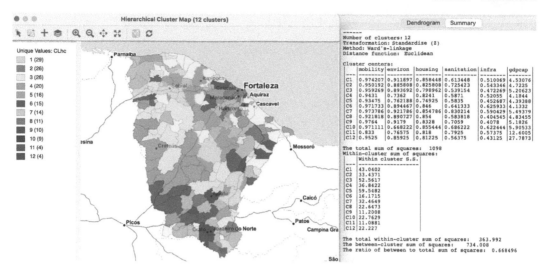

Figure 12.2: Hierarchical Clustering – Ward's method, Ceará

Method	WSS	BSS/TSS	Entropy	Simpson	Join Count
Hierarchical	363.992	0.6685	2.3389	0.1065	0.2120
K-Means	349.122	0.6820	2.3209	0.1093	0.2260
SCHC	588.391	0.4641	1.6130	0.2758	0.6780
SKATER	623.089	0.4325	1.3353	0.3963	0.7940
REDCAP	589.912	0.4627	1.5973	0.2772	0.6620
AZP	618.341	0.4368	1.9864	0.1884	0.5300
AZP-Initial	545.063	0.5036	1.5899	0.2909	0.5640
Maxp-12	675.707	0.3846	2.3564	0.1018	0.5460

Figure 12.3: Internal Validation Measures

Since the hierarchical cluster outcome is not spatially constrained, the fragmentation characteristics are also listed for each of the 12 clusters individually. The size is given (N), its share in the total number of observations (Fraction), the number of sub-clusters (#Sub), raw and standardized Entropy and Simpson index, as well as the minimum, maximum and mean size of the subclusters. This provides a detailed picture of the degree of fragmentation by cluster. For example, the table shows that cluster 11 consists of 4 compact observations (fragmentation results given as 0), whereas cluster 12 is made up of 4 singletons, the most fragmented result (yielding a standardized value of 1 for both entropy and Simpson's index). The best result, in the sense of the least fragmentation or least diverse, is obtained for cluster 1. Its 29 observations are divided among 9 subclusters (smallest standardized entropy, i.e., worst diversity, of 0.797 and largest standardized Simpson of 2.258).

In addition to the fragmentation measures, the join count ratio is computed for each individual cluster as well. This provides the number of neighbors and the count of joins that belong to the same cluster, yielding the join count ratio. At the bottom, the overall join count ratio is listed. Taking into account the neighbor structure yields cluster 11 (which has no subclusters, hence is compact) with the highest score of 0.476, closely followed by cluster 1 with 0.405.

Since the result pertains to a method that is not spatially constrained, there are no measures for compactness and diameter.

Comparing the overall results in Figure 12.3 confirms the superiority of the K-Means outcome in terms of fit, with the best WSS and BSS/TSS. In general, the unconstrained clustering

```
Spatial Validation
Variable: CLhc

Fragmentation:
|#Clusters|Entropy|Entropy*|Simpson|Simpson*|
|---------|-------|--------|-------|--------|
|12       |2.3389 |0.9413  |0.1065 |1.2774  |

Subcluster Fragmentation:
|Cluster|N |Fraction|#Sub|Entropy|Entropy*|Simpson|Simpson*|Min|Max|Mean|
|-------|--|--------|----|-------|--------|-------|--------|---|---|----|
|1      |29|0.1576  |9   |1.7512 |0.7970  |0.2509 |2.2580  |1  |13 |3   |
|10     |9 |0.0489  |7   |1.8310 |0.9410  |0.1852 |1.2963  |1  |3  |1   |
|11     |4 |0.0489  |0   |0      |0       |0      |0       |0  |0  |0   |
|12     |4 |0.0217  |4   |1.3863 |1       |0.2500 |1       |1  |1  |1   |
|2      |26|0.1413  |14  |2.4258 |0.9192  |0.1095 |1.5325  |1  |5  |1   |
|3      |26|0.1413  |14  |2.3979 |0.9086  |0.1154 |1.6154  |1  |6  |1   |
|4      |20|0.1087  |11  |2.2503 |0.9384  |0.1200 |1.3200  |1  |4  |1   |
|5      |16|0.0870  |9   |1.9231 |0.8752  |0.1875 |1.6875  |1  |5  |1   |
|6      |15|0.0815  |8   |1.9338 |0.9300  |0.1644 |1.3156  |1  |4  |1   |
|7      |14|0.0761  |10  |2.1682 |0.9416  |0.1327 |1.3265  |1  |3  |1   |
|8      |11|0.0598  |9   |2.1458 |0.9766  |0.1240 |1.1157  |1  |2  |1   |
|9      |10|0.0543  |6   |1.6094 |0.8982  |0.2400 |1.4400  |1  |4  |1   |

Join Count Ratio
|Cluster|N  |Neighbors|Join Count|Ratio |
|-------|---|---------|----------|------|
|1      |29 |158      |64        |0.4051|
|10     |9  |47       |6         |0.1277|
|11     |4  |21       |10        |0.4762|
|12     |4  |18       |0         |0     |
|2      |26 |158      |30        |0.1899|
|3      |26 |134      |26        |0.1940|
|4      |20 |116      |18        |0.1552|
|5      |16 |96       |20        |0.2083|
|6      |15 |83       |18        |0.2169|
|7      |14 |80       |8         |0.1000|
|8      |11 |43       |4         |0.0930|
|9      |10 |46       |8         |0.1739|
|All    |184|1000     |212       |0.2120|

Compactness:
N/A: clusters are not spatially constrained.

Diameter:
N/A: clusters are not spatially constrained.
```

Figure 12.4: Internal Validation Result – Hierarchical Clustering

methods do (much) better on these criteria than the spatially constrained results, with only AZP-initial coming somewhat close. This matches a similar dichotomy for the fragmentation indicators, with the spatially constrained outcomes much less equally balanced (smaller entropy, larger Simpson) than the classical results. Interestingly, the minimum population size imposed in max-p yields a more balanced outcome, with entropy and Simpson on a par with the classical results (but much worse overall fit). Finally, the overall join count ratio confirms the superiority of the spatially constrained results in this respect, with SKATER yielding the highest ratio.

For a spatially constrained clustering method, the results window for **Validation** is slightly different, as illustrated in Figure 12.5 for AZP with a SCHC initial region.

The fragmentation summary takes the same form as before, but there is no report on subcluster fragmentation. The join count ratio is again included, with the outcome for all individual clusters as well as the overall ratio. In the example, the highest ratio is obtained in cluster 1, with a value of 0.6759, compared to the overall ratio of 0.564.

Two additional summary tables include the computation of the IPQ as a compactness index, as well as the diameter based on the spatial weights in each cluster.

The highest values for IPQ are obtained for the singletons, which is not very informative. Of the four largest clusters, cluster 4 (with 14 observations) seems the most compact, with a ratio of 0.032. Overall, the smaller clusters clearly do better on this criterion. For example, cluster 5 (with 5 observations) achieves a ratio of 0.066, and cluster 7 (with 4 observations) obtains a ratio of 0.128. For comparison, the largest ratio obtained for a singleton is for cluster 12, with a ratio of 0.539.

Finally, the diameter ranges from 0 (for singletons) to 23 for cluster 2 (with 51 observations). Standardized for the number of observations in each cluster, cluster 1 is the most connected

```
Spatial Validation
Variable: CLazhsc

Fragmentation:
|#Clusters|Entropy|Entropy*|Simpson|Simpson*|
|---------|-------|--------|-------|--------|
|12       |1.5899 |0.6398  |0.2909 |3.4913  |

Subcluster Fragmentation:
N/A: clusters are spatially constrained.

Join Count Ratio
|Cluster|N  |Neighbors|Join Count|Ratio |
|-------|---|---------|----------|------|
|1      |82 |432      |292       |0.6759|
|10     |1  |7        |0         |0     |
|11     |1  |3        |0         |0     |
|12     |1  |3        |0         |0     |
|2      |51 |278      |148       |0.5324|
|3      |16 |96       |50        |0.5208|
|4      |14 |78       |44        |0.5641|
|5      |5  |31       |8         |0.2581|
|6      |4  |21       |10        |0.4762|
|7      |4  |17       |6         |0.3529|
|8      |3  |22       |4         |0.1818|
|9      |2  |12       |2         |0.1667|
|All    |184|1000     |564       |0.5640|

Compactness:
|Cluster|Area    |Perimeter    |IPQ     |
|-------|--------|-------------|--------|
|1      |8.189e+10|1.427e+07   |0.005054|
|10     |8.449e+08|173988.7234 |0.350738|
|11     |7.908e+07|50568.7732  |0.388616|
|12     |6.172e+08|119924.2678 |0.539283|
|2      |3.282e+10|7359182.0926|0.007615|
|3      |1.081e+10|2309823.5248|0.025457|
|4      |1.139e+10|2122643.1187|0.031759|
|5      |4.169e+09|893856.1676 |0.065573|
|6      |1.78e+09 |424006.8279 |0.124386|
|7      |3.086e+09|549912.0216 |0.128238|
|8      |9.969e+08|316856.7893 |0.124782|
|9      |1.475e+09|322935.7613 |0.177718|

Diameter:
|Cluster|Steps|Ratio   |
|-------|-----|--------|
|1      |22   |0.268293|
|10     |0    |0.000000|
|11     |0    |0.000000|
|12     |0    |0.000000|
|2      |23   |0.450980|
|3      |5    |0.312500|
|4      |7    |0.500000|
|5      |3    |0.600000|
|6      |2    |0.500000|
|7      |3    |0.750000|
|8      |2    |0.666667|
|9      |1    |0.500000|
```

Figure 12.5: Internal Validation Result – AZP with Initial Region

(22 steps with 82 observations, for a ratio of 0.286), closely followed by cluster 3, which is much smaller (16 observations with 5 steps, for a ratio of 0.3125).

Clearly, the different dimensions of performance highlight distinct characteristics of each cluster. In each particular application, some dimensions may be more relevant than others, requiring a careful assessment.

12.3   External Validity

External validity measures are designed to compare a clustering result to an underlying *truth*, typically carried out through experiments on artificial data sets (see, e.g., Meila, 2015; Aydin et al., 2021). However, such indices can also be employed to assess how much different clustering methods result in classifications that are similar in some respects. In general, the clusterings may be for different number of clusters, but for simplicity, only the case where k is the same in both classifications is taken into account.

	Hierarchical	K-Means	SCHC	SKATER	REDCAP	AZP	AZP_init	Maxp-12
Hierarchical	1.000	0.352	0.129	0.075	0.144	0.150	0.162	0.132
K-Means		1.000	0.144	0.062	0.155	0.158	0.161	0.136
SCHC			1.000	0.411	0.910	0.457	0.677	0.241
SKATER				1.000	0.374	0.232	0.354	0.185
REDCAP					1.000	0.471	0.700	0.251
AZP						1.000	0.558	0.355
AZP-Initial							1.000	0.251
Maxp-12								1.000

Figure 12.6: Adjusted Rand Index

Two categories of measures are considered here. First, two classic measures of external validity are considered, i.e., the *Adjusted Rand Index* (ARI) and the *Normalized Information Distance* (NID). This is followed by some examples of how the linking and brushing capabilities in GeoDa can be used to gain insight into the overlap between clusters, including by means of the specialized *cluster match map*.

12.3.1 Classic Measures

12.3.1.1 Adjusted Rand Index

Classic external validity measures can be categorized into pair counting-based methods and information-theoretic measures. Arguably the most commonly used index based on the former is the *Adjusted Rand Index* (ARI) originally suggested by Hubert and Arabie (1985).

The ARI is derived by classifying the number of pairs of observations in terms of the extent to which they belong to common clusters. More precisely, all pairs of observations in two clusterings are considered, e.g., clusters labeled P and V. The pairs are classified into four categories: N_{11}, the number of pairs in the same cluster in both classifications; N_{00} the number of pairs in different clusters in both classifications; N_{01} the pairs in the same cluster in P, but in different clusters in V; and N_{10}, the pairs that are in different clusters in P, but in the same cluster in V.

The Adjusted Rand Index (ARI) is then defined as:

$$ARI(P,V) = \frac{2N_{00}N_{11} - N_{01}N_{10}}{(N_{00} + N_{01})(N_{01} + N_{11}) + (N_{00} + N_{10})(N_{10} + N_{11})}$$

This index is calibrated such that the maximum value is 1. The measure is symmetric, so that the ARI of P relative to V is the same as that of V relative to P.

For the Ceará example, the results of each pairwise ARI are listed in Figure 12.6. There is a clear split between the match of the nonspatial methods and that of the spatially constrained clusters. The former yields an ARI of 0.352, more than double the value for any nonspatial/spatial pair. The spatially constrained clusters are more similar among themselves, with the closest match between SCHC and REDCAP for an ARI of 0.910. Other close matches are between REDCAP and AZP-init (0.700), and SCHC and AZP-init (0.677). The lowest values are between the nonspatial clusters and SKATER (respectively, 0.062 for K-Means and 0.075 for Hierarchical clustering).

12.3.1.2 Normalized Information Distance

A second class of methods is based on information theoretic concepts, such as the entropy and mutual information (Vinh et al., 2010). The entropy is defined above as a measure of the balance in a given cluster result (see Section 12.2.2). A second important concept is the

	Hierarchical	K-Means	SCHC	SKATER	REDCAP	AZP	AZP_init	Maxp-12
Hierarchical	0.000	0.389	0.707	0.755	0.701	0.631	0.653	0.640
K-Means		0.000	0.678	0.748	0.673	0.628	0.652	0.623
SCHC			0.000	0.461	0.108	0.424	0.287	0.552
SKATER				0.000	0.508	0.588	0.497	0.589
REDCAP					0.000	0.421	0.267	0.559
AZP						0.000	0.387	0.402
AZP-Initial							0.000	0.530
Maxp-12								0.000

Figure 12.7: Normalized Information Distance

mutual information between clusters P and V. It is computed as:

$$I(P,V) = \sum_{i=1}^{k} \sum_{j=1}^{k} \frac{n_{ij}}{n} \log \frac{n_{ij}/n}{n_i n_j / n^2},$$

where n_{ij} is the number of observation pairs that are in cluster i in P and in cluster j in V. The Normalized Information Distance is then obtained as:

$$NID = 1 - \frac{I(P,V)}{\max(H(P), H(V))},$$

where H is the entropy of a given clustering. The measure is such that lower values suggest a better fit (the minimum is zero).

The results for the Ceará example are shown in Figure 12.7. The relative ranking is essentially the same in qualitative terms as for ARI, with only marginal differences. As before, the best match is between SCHC and REDCAP, and the worst fit is between the nonspatial methods and SKATER.

12.3.2   Visualizing Cluster Match

A more informal way to assess the match between cluster results can be obtained by means of the linking and brushing inherent in the mapping functionality. The point of departure is that each cluster classification corresponds to a categorical variable that can be used to create a *unique values map*.

12.3.2.1   Linking Cluster Maps

With the maps side by side for two clusterings, a category can be selected in one and the matching observations identified in the other. For example, in Figure 12.8, the cluster maps for K-Means and SKATER are shown side-by-side, two clusterings that showed the least correspondence according to ARI and NID. With cluster 1 selected in the K-Means map (the pointer is over the first box in the legend), its 31 observations are highlighted in both maps. One can *visually* identify the match between the two classifications by locating pairs of observations from cluster 1 for K-Means that are in the same cluster in both maps, and those that are not. For example, the pair highlighted with the blue rectangle is in the same cluster in both classifications (this happens to be labeled cluster 1 in both maps, but that is pure coincidence). In contrast, the pair highlighted with the red rectangle no longer belongs to the same cluster for SKATER (one observation is now in cluster 3 and one in cluster 6).

In addition to considering pairs of observations, which follows the pair counting logic, one can also exploit the linking functionality to identify the overlap between clusters as such. For example, in Figure 12.9, the cluster maps are shown for SCHC and REDCAP, the two

Figure 12.8: K-Means and SKATER overlap

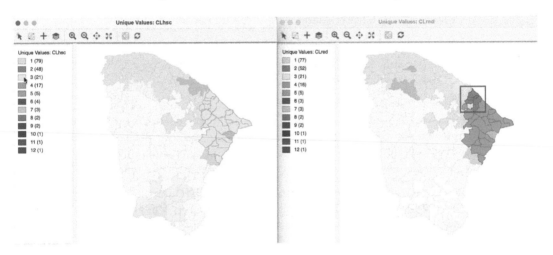

Figure 12.9: SCHC and REDCAP overlap

clusterings that had the closest match using the external validity indices. The cluster maps indeed are highly similar. For example, of the 21 observations in cluster 3 for SCHC, 16 make up cluster 4 for REDCAP. The five remaining observations (highlighted in the red rectangle) are part of cluster 2 for REDCAP (which has 52 observations).

12.3.2.2 Cluster Match Map

A slightly more automatic approach to visualizing the overlap between different cluster categories is offered by the *Cluster Match Map*. This function is invoked as **Cluster >** **Cluster Match Map** from the menu, or from the cluster toolbar icon drop-down list, as shown in Figure 12.1.

The **Cluster Match Map Settings** dialog requires the selection of the **Origin Cluster Indicator** and **Target Cluster Indicator**. For example, the first cluster could be SKATER, with indicator **CLskat**, and the second cluster K-Means, with indicator **CLkm** (the reverse of the situation in Figure 12.8). The cluster to compare in the origin cluster is selected by

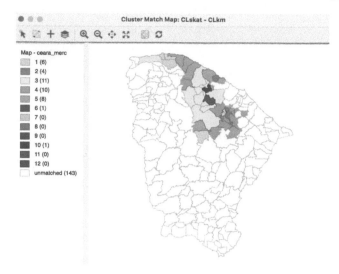

Figure 12.10: Cluster Match Map – SKATER and K-MEANS

checking the corresponding box. The resulting cluster overlap is by default stored in the variable **CL_COM** (as usual, this can be changed).

The result for cluster 2 in SKATER (41 observations) is illustrated in Figure 12.10. This is the same outcome as the linked observations in the K-Means cluster map that result from selecting cluster 2 in the SKATER cluster map, but the number of observations in each overlap is listed as well. The cluster label is the label from the target cluster, i.e., K-Means. In addition, the color codes match the codes in the K-Means cluster map classification.

In the example, the 41 observations in cluster 2 of SKATER map into 6 observations in K-Means cluster 1, 4 in cluster 2, 11 in cluster 3, 10 in cluster 4, 8 in cluster 5 and one each in clusters 6 and 10. The 143 observations listed as **unmatched** did not belong to cluster 2 for SKATER (184-41=143).

In sum, rather than having to rely on a visual assessment of the overlap between clusters, the *Cluster Match Map* provides a simple numeric summary.

12.4   Beyond Clustering

The combination of dimension reduction in the variable space and the grouping of observations into clusters can provide very powerful insights into the structure of multivariate data. However, similar to the outcomes of a data exploration discussed in Volume 1, the ultimate results do not provide explanations, nor do they suggest particular spatial processes. In essence, a clustering exercise results in a *simplification* of the original complex data structure in order to facilitate further analysis. As pointed out in this volume in several places, any such clustering exercise needs to be carried out with caution. A mechanical approach that relies on the default values in the software must be avoided.

Of course, this leaves the question of how to select the proper combination of algorithms, parameters and other tuning factors. Unfortunately, there is no hard and fast answer to address these choices. To some extent, it depends on the goal of the analysis. For example, sometimes it is required to have spatially contiguous observations in each cluster, e.g., as a legal requirement for electoral redistricting purposes or for solutions to a location-allocation problem. Obviously, in such instances, the classic nonspatial clustering techniques fall short. On the other hand, when there is no such strict requirement, the loss in *fit* due to the *constrained* spatial solution is something that needs to be carefully evaluated in each particular instance. It should be kept in mind that unconstrained solutions always obtain a better fit than constrained ones, so a sole focus on fit will favor nonspatial approaches.

There is no general guidance in the choice between spatial or a-spatial solutions. A case in point is the delineation of so-called *housing submarkets* in urban studies, i.e., clusters of similar housing units used to explain variations in their value. Part of the literature argues that the main objective should be to maximize the similarity of housing units in a given submarket, whereas another part of the literature insists on imposing a spatial contiguity constraint (for an extensive discussion, see Anselin and Amaral, 2023).

Sometimes, the construction of clusters is an objective in itself, such as in the redistricting example mentioned earlier. However, many times the goal is to simplify the complex multivariate structure of the data to provide a starting point for further analysis. For example, this is the case when statistical or privacy concerns require a minimum size for the denominator in rates, such as for rates of rare diseases. A common solution is to carry out a constrained spatial clustering exercise, which guarantees a minimum population at risk. As the several examples covered illustrate, such an approach typically does not yield a unique solution. Again, careful sensitivity analyses are needed to assess how the various design decisions affect the ultimate conclusions (e.g., about the location of hot spots or cold spots).

In the spatial analysis literature, clustering is often seen as a way to address the *modifiable areal unit problem* or MAUP (see Chapter 21 in Volume 1). When elemental units are available (e.g., individual housing units in the housing market example), a clustering approach allows for the grouping of observations according to certain objective functions. To some extent, this avoids the arbitrariness of the scale and spatial arrangement of administrative aggregate units, but it also introduces a degree of uncertainty related to the choice of algorithm, tuning parameters, etc., mentioned before.

In a similar vein, clustering methods can be used to address the problem of *spatial heterogeneity* in spatial regression analysis (Anselin and Rey, 2014). Such heterogeneity pertains to structural breaks in the data that result in separate models and/or separate model coefficients for spatial subsets of the data, referred to as *spatial regimes*. The grouping of observations into regimes is a particular application of spatially constrained clustering. Tight integration between the estimation of model parameters and the delineation of spatial regimes is achieved in models of *endogenous spatial regimes*, a subject of active ongoing research (Anselin and Amaral, 2023).

Bibliography

Akhanli, S. E. and Hennig, C. (2020). Comparing clusterings and numbers of clusters by aggregation of calibrated clustering validity indexes. *Statistics and Computing*, 30: 1523–1544.

Algeri, C., Anselin, L., Forgione, A. F., and Migliardo, C. (2022). Spatial dependence in the technical efficiency of local banks. *Papers in Regional Science*, 101:385–416.

Amaral, P., de Carvalho, L. R., Rocha, T. A. H., da Silva, N. C., and Vissoci, J. R. N. (2019). Geospatial modeling of microcephaly and zika virus spread patterns in Brazil. *PLoS ONE*, 14.

Anselin, L. and Amaral, P. (2023). Endogenous spatial regimes. *Journal of Geographical Systems*. https://doi.org/10.1007/s10109-023-00411-2

Anselin, L. and Rey, S. J. (2014). *Modern Spatial Econometrics in Practice, A Guide to GeoDa, GeoDaSpace and PySAL*. GeoDa Press, Chicago, IL.

Assunção, R. M., Neves, M., Câmara, G., and Da Costa Freitas, C. (2006). Efficient regionalization techniques for socio-economic geographical units using minimum spanning trees. *International Journal of Geographical Information Science*, 20:797–811.

Arthur, D. and Vassilvitskii, S. (2007). k-means++: the advantages of careful seeding. In Gabow, H., editor, *SODA 07, Proceedings of the Eighteenth Annual ACM-SIAM Symposium on Discrete Algorithms*, pages 1027–1035, Philadelphia, PA. Society for Industrial and Applied Mathematics.

Aydin, O., Janikas, M. V., Assunção, R. M., and Lee, T.-H. (2018). SKATER-CON: unsupervised regionalization via stochastic tree partitioning within a consensus framework using random spanning trees. In *2nd ACM SIGSPATIAL International Workshop on AI for Geographic Knowledge Discovery (GeoAI'18)*, Seattle, WA.

Aydin, O., Janikas, M. V., Assunção, R. M., and Lee, T.-H. (2021). A quantitative comparison of regionalization methods. *International Journal of Geographical Information Science*, 35:2287–2315.

Banerjee, S. and Roy, A. (2014). *Linear Algebra and Matrix Analysis for Statistics*. Chapman & Hall/CRC, Boca Raton.

Barnes, J. and Hut, P. (1986). A hierarchical O(N log N) force-calculation algorithm. *Nature*, 324:446–449.

Borg, I. and Groenen, P. J. (2005). *Modern Multidimensional Scaling, Theory and Applications (2nd Ed)*. Springer, New York, NY.

Chavent, M., Kuentz-Simonet, V., Labenne, A., and Saracco, J. (2018). ClustGeo: an R package for hierarchical clustering with spatial constraints. *Computational Statistics*, 33:1799–1822.

Cheruvelil, K. S., Yuan, S., Webster, K. E., Tan, P.-N., Jean-François, Collins, S. M., Fergus, C. E., Scott, C. E., Henry, E. N., Soranno, P., Filstrup, C. T., and Wagner, T. (2017). Creating multithemed ecological regions for macroscale ecology: testing a flexible, repeatable, and accessible clustering method. *Ecology and Evolution*, 7:3046–3058.

Church, R. L. and Murray, A. T. (2009). *Business Site Selection, Location Analysis, and GIS*. John Wiley and Sons, Hoboken, NJ.

Cova, T. J. and Church, R. L. (2000). Contiguity constraints for single-region site search problems. *Geographical Analysis*, 32:306–329.

Cressie, N. and Wikle, C. K. (2011). *Statistics for Spatio-Temporal Data*. John Wiley and Sons, Hoboken, NJ.

de Hoon, M., Imoto, S., and Miyano, S. (2017). The C clustering library. Technical report, The University of Tokyo, Institute of Medical Science, Human Genome Center, Tokyo, Japan.

de Leeuw, J. (1977). Applications of convex analysis to multidimensional scaling. In Barra, J., Brodeau, F., Romier, G., and van Cutsem, B., editors, *Recent Developments in Statistics*, pages 133–145. North Holland, Amsterdam.

de Leeuw, J. and Mair, P. (2009). Multidimensional scaling using majorization: SMACOF in R. *Journal of Statistical Software*, 21.

Dempster, A., Laird, N., and Rubin, D. (1977). Maximum likelihood from incomplete data via the EM algorithm. *Journal of the Royal Statistical Society, Series B*, 39:1–38.

Dray, S. and Jombart, T. (2011). Revisiting Guerry's data: introducing spatial constraints in multivariate analysis. *The Annals of Applied Statistics*, 5(4):2278–2299.

Dray, S., Saïd, S., and Débias, F. (2008). Spatial ordination of vegetation data using a generalization of Wartenberg's multivariate spatial correlation. *Journal of Vegetation Science*, 19:45–56.

Duque, J. C., Anselin, L., and Rey, S. J. (2012). The max-p-regions problem. *Journal of Regional Science*, 52:397–419.

Duque, J. C., Church, R. L., and Middleton, R. S. (2011). The p-regions problem. *Geographical Analysis*, 43:104–126.

Duque, J. C., Ramos, R., and Suriñach, J. (2007). Supervised regionalization methods: a survey. *International Regional Science Review*, 30:195–220.

Everitt, B. S., Landau, S., Leese, M., and Stahl, D. (2011). *Cluster Analysis, 5th Edition*. John Wiley, New York, NY.

Feng, X., Rey, S. J., and Wei, R. (2022). The max-p-compact-regions problem. *Transactions in GIS*, 26:717–734.

Fränti, P. and Sieranoja, S. (2018). K-means properties on six clustering benchmark datasets. *Applied Intelligence*, 48:4743–4759.

Frichot, E., Schoville, S., Bouchard, G., and François, O. (2012). Correcting principal component maps for effects of spatial autocorrelation in population genetic data. *Frontiers in Genetics*, 3:11–9.

Gelman, A. and Loken, E. (2014). The statistical crisis in science. *American Scientist*, 102:460–465.

Glover, F. (1977). Heuristics for integer programming using surrogate constraints. *Decision Science*, 8:156–166.

Goodall, W. (1954). Objective methods for the classification of vegetation III. An essay on the use of factor analysis. *Australian Journal of Botany*, 2:304–324.

Gordon, A. (1996). A survey of constrained classification. *Computational Statistics and Data Analysis*, 21:17–29.

Guo, D. (2008). Regionalization with dynamically constrained agglomerative clustering and partitioning (REDCAP). *International Journal of Geographical Information Science*, 22:801–823.

Guo, D. (2009). Greedy optimization for contiguity-constrained hierarchical clustering. In *2013 IEEE 13th International Conference on Data Mining Workshops*, pages 591–596, Los Alamitos, CA, USA. IEEE Computer Society.

Guo, D. and Wang, H. (2011). Automatic region building for spatial analysis. *Transactions in GIS*, 15:29–45.

Haining, R. F., Wise, S., and Ma, J. (2000). Designing and implementing software for spatial statistical analysis in a GIS environment. *Journal of Geographical Systems*, 2(3):257–286.

Halkidi, M., Vazirgiannis, M., and Hennig, C. (2015). Method-independent indices for cluster validation and estimating the number of clusters. In Hennig, C., Meila, M., Murtagh, F., and Rocci, R., editors, *Handbook of Cluster Analysis*, pages 596–618. CRC Press, Boca Raton, FL.

Han, J., Kamber, M., and Pei, J. (2012). *Data Mining (Third Edition)*. MorganKaufman, Amsterdam.

Hartigan, J. A. (1972). Direct clustering of a data matrix. *Journal of the American Statistical Association*, 67:123–129.

Hartigan, J. A. (1975). *Clustering Algorithms*. John Wiley, New York, NY.

Hartigan, J. A. and Wong, M. A. (1979). Algorithm AS 136: A k-means clustering algorithm. *Applied Statistics*, 28:100–108.

Hastie, T., Tibshirani, R., and Friedman, J. (2009). *The Elements of Statistical Learning (2nd Edition)*. Springer, New York, NY.

Hinton, G. E. and Roweis, S. T. (2003). Stochastic neighbor embedding. In Becker, S., Thun, S., and Obermayer, K., editors, *Advances in Neural Information Processing Systems 15 (NIPS 2002)*, pages 833–840, Vancouver, BC. NIPS.

Hotelling, H. (1933). Analysis of a complex of statistical variables into principal components. *Journal of Educational Psychology*, 24:417–441.

Hubert, L. J. and Arabie, P. (1985). Comparing partitions. *Journal of Classification*, 2:193–218.

Jain, A. K. (2010). Data clustering: 50 years beyond k-means. *Pattern Recognition Letters*, 31:651–666.

Jain, A. K. and Dubes, R. C. (1988). *Algorithms for Clustering Data*. Prentice Hall, Englewood Cliffs, NJ.

James, G., Witten, D., Hastie, T., and Tibshirani, R. (2013). *An Introduction to Statistical Learning, with Applications in R*. Springer-Verlag, New York, NY.

Jombart, T., Devillard, S., Dufour, A.-B., and Pontier, D. (2008). Revealing cryptic spatial patterns in genetic variability by a new multivariate method. *Heredity*, 101:92–103.

Kaiser, H. F. (1960). The application of electronic computers to factor analysis. *Educational and Psychological Measurement*, 20:141–151.

Kaufman, L. and Rousseeuw, P. (2005). *Finding Groups in Data: An Introduction to Cluster Analysis*. John Wiley, New York, NY.

Kolak, M., Bhatt, J., Park, Y. H., Padrón, N. A., and Molefe, A. (2020). Quantification of neighborhood-level social determinants of health in the continental united states. *JAMA Network Open*, 3(1):e1919928–e1919928.

Kruskal, J. (1964). Multidimensional scaling by optimizing goodness of fit to a non-metric hypothesis. *Psychometrika*, 29:1–17.

Lankford, P. M. (1969). Regionalization: theory and alternative algorithms. *Geographical Analysis*, 1:196–212.

Laura, J., Li, W., Rey, S. J., and Anselin, L. (2015). Parallelization of a regionalization heuristic in distributed computing platforms – a case study of parallel-p-compact-regions problem. *International Journal of Geographical Information Science*, 29:536–555.

Lee, J. A. and Verleysen, M. (2007). *Nonlinear Dimensionality Reduction*. Springer-Verlag, New York, NY.

Li, W., Church, R., and Goodchild, M. F. (2014). The p-compact-regions problem. *Geographical Analysis*, 46:250–273.

Lloyd, S. P. (1982). Least squares quantization in PCM. *IEEE Transactions on Information Theory*, 28:129–136.

Mead, A. (1992). Review of the development of multidimensional scaling methods. *Journal of the Royal Statistical Society. Series D (The Statistician)*, 41:27–39.

Meila, M. (2015). Criteria for comparing clusterings. In Hennig, C., Meila, M., Murtagh, F., and Rocci, R., editors, *Handbook of Cluster Analysis*, pages 619–635. CRC Press, Boca Raton, FL.

Metropolis, N., Rosenbluth, A., Rosenbluth, M., Teller, A., and Teller, E. (1953). Equations for state calculations by fast computing machines. *Journal of Chemical Physics*, 21:1087–1092.

Müllner, D. (2011). Modern hierarchical, agglomerative clustering algorithms. *ArXiv:1109.2378[stat.ML]*.

Murray, A. T. and Grubesic, T. H. (2002). Identifying non-hierarchical spatial clusters. *International Journal of Industrial Engineering*, 9:86–95.

Murray, A. T. and Shyy, T.-K. (2000). Integrating attribute and space characteristics in choropleth display and spatial data mining. *International Journal of Geographical Information Science*, 14:649–667.

Murtagh, F. (1985). A survey of algorithms for contiguity-constrained clustering and related problems. *The Computer Journal*, 28:82–88.

Newman, M. (2018). *Networks*. Oxford University Press, Oxford, United Kingdom.

Ng, R. R. and Han, J. (2002). CLARANS: a method for clustering objects for spatial data mining. *IEEE Transactions on Knowledge and Data Engineering*, 14:1003–1016.

Niemi, R., Grofman, B., Carlucci, C., and Hofeller, T. (1990). Measuring compactness and the role of a compactness standard in a test for partisan and racial gerrymandering. *The Journal of Politics*, 52:1155–1181.

Openshaw, S. (1973). A regionalisation program for large data sets. *Computer Applications*, 3-4:136–147.

Openshaw, S. (1977). A geographical solution to scale and aggregation problems in region-building, partitioning and spatial modeling. *Transactions of the Institute of British Geographers*, 2:459–472.

Openshaw, S. and Rao, L. (1995). Algorithms for reengineering the 1991 census geography. *Environment and Planning A*, 27:425–446.

Padilha, V. A. and Campello, R. J. (2017). A systematic comparative evaluation of biclustering techniques. *BMC Bioinformatics*, 18:55.

Pearson, K. (1901). On lines and planes of closest fit to systems of points in space. *Philosophical Magazine*, 2:559–572.

Polsby, D. and Popper, R. (1990). The third criterion: compactness as a procedural safeguard against partisan gerrymandering. *Yale Law and Policy Review*, 9:301–353.

Recchia, A. (2010). Contiguity-constrained hierarchical agglomerative clustering using SAS. *Journal of Statistical Software*, 22.

Saxon, J. (2020). Reviving legislative avenues for gerrymandering reform with a flexible, automated tool. *Political Analysis*, 28:372–394.

Schubert, E. and Rousseeuw, P. J. (2019). Faster k-Medoids clustering: improving the PAM, CLARA, and CLARANS algorithms. In Amato, G., Gennaro, C., Oria, V., and Radovanović, M., editors, *Similarity Search and Applications, SISAP 2019*, pages 171–187, Cham, Switzerland. Springer Nature.

She, B., Duque, J. C., and Ye, X. (2017). The network-max-p-regions model. *International Journal of Geographical Information Science*, 31:962–981.

Shepard, R. N. (1962a). The analysis of proximities: multidimensional scaling with an unknown distance function I. *Psychometrika*, 27:125–140.

Shepard, R. N. (1962b). The analysis of proximities: multidimensional scaling with an unknown distance function II. *Psychometrika*, 27:219–246.

Shi, J. and Malik, J. (2000). Normalized cuts and image segmentation. *IEEE Transactions on Pattern Analysis and Machine Intelligence*, 22:888–905.

Simpson, E. H. (1949). Measurement of diversity. *Nature*, 163:688.

Tanay, A., Sharan, R., and Shamir, R. (2004). Biclustering algorithms: a survey. In Aluru, S., editor, *Handbook of Computational Molecular Biology*, pages 26–1–17. Chapman & Hall/CRC, Boca Raton, FL.

Teixeira, L. V., Assunção, R. M., and Loschi, R. H. (2015). A generative spatial clustering model for random data through spanning trees. In *2015 IEEE International Conference on Data Mining*, pages 997–1026, Los Alamitos, CA, USA. IEEE Computer Society.

Tibshirani, R., Walther, G., and Hastie, T. (2001). Estimating the number of clusters in a data set via the gap statistic. *Journal of the Royal Statistical Society, Series B*, 63:411–423.

Torgerson, W. S. (1952). Multidimensional scaling, I: theory and method. *Psychometrika*, 17:401–419.

Torgerson, W. S. (1958). *Theory and Methods of Scaling*. John Wiley, New York, NY.

van der Maaten, L. (2014). Accelerating t-SNE using tree-based algorithms. *Journal of Machine Learning Research*, 15:1–21.

van der Maaten, L. and Hinton, G. (2008). Visualizing data using t-SNE. *Journal of Machine Learning Research*, 9:2579–2605.

Vinh, N. X., Epps, J., and Bailey, J. (2010). Information theoretic measures for clustering comparison: variants, properties, normalization and correction for chance. *Journal of Machine Learning Research*, 11:2837–2854.

von Luxburg, U. (2007). A tutorial on spectral clustering. *Statistical Computing*, 27:395–416.

Ward, J. H. (1963). Hierarchical grouping to optimize an objective function. *Journal of the American Statistical Association*, 58:236–244.

Wattenberg, M., Viégas, F., and Johnson, I. (2016). How to use t-SNE effectively. *Distill*. http://doi.org/10.23915/distill.00002

Webster, R. and Burrough, P. (1972). Computer-based soil mapping of small areas from sample data II classification smoothing. *Journal of Soil Science*, 23:222–234.

Wei, R., Rey, S., and Knaap, E. (2021). Efficient regionalization for spatially explicit neighborhood delineation. *International Journal of Geographical Information Science*, 31:135–151.

Wikle, C. K., Zammit-Mangion, A., and Cressie, N. (2019). *Spatio-Temporal Statistics with R*. CRC Press, Boca Raton, FL.

Wise, S., Haining, R., and Ma, J. (1997). Regionalisation tools for exploratory spatial analysis of health data. In Fischer, M. M. and Getis, A., editors, *Recent Developments in Spatial Analysis: Spatial Statistics, Behavioural Modelling, and Computational Intelligence*, pages 83–100, New York, NY. Springer.

Wolf, L. J., Knaap, E., and Rey, S. (2021). Geosilhouettes: geographical measures of cluster fit. *EPB, Urban Analytics and City Science*, 48:521–539.

Yianilos, P. N. (1993). Data structures and algorithms for nearest neighbor search in general metric spaces. In *Proceedings of the ACM-SIAM Symposium on Discrete Algorithms*, pages 311–321, Philadelphia, PA. SIAM.

Yuan, S., Tan, P.-N., Cheruvelil, K. S., Collins, S. M., and Soranno, P. A. (2015). Constrained spectral clustering for regionalization: exploring the trade-off between spatial contiguity and landscape homogeneity. In *2015 IEEE International Conference on Data Science and Advanced Analytics (DSAA)*, Paris, France.

Index

Note: Locators in *italics* represent figures. Locators followed by n and number represent footnote and note number respectively.

Milton Keynes UK
Ingram Content Group UK Ltd.
UKHW020846141024
449569UK00003B/87